RED CLOUD

SOUTH
DAKOTA
BIOGRAPHY
SERIES

Laura Ingalls Wilder
by Pamela Smith Hill

Wild Bill Hickok and Calamity Jane
by James D. McLaird

Seth Bullock
by David A. Wolff

Red Cloud
by John D. McDermott

RED CLOUD

OGLALA LEGEND

JOHN D. MCDERMOTT

South Dakota
Historical Society
Press *Pierre*

Red Cloud is Volume 4 in the South Dakota Biography Series.

This publication is funded, in part, by the
Great Plains Education Foundation, Inc., Aberdeen, S.Dak.

Library of Congress Cataloging-in-Publication data
McDermott, John D. (John Dishon)
Red Cloud : Oglala legend / John D. McDermott.
    pages cm. — (South Dakota biography series; volume 4)
Includes bibliographical references and index.
ISBN 978-1-941813-02-7 (alk. paper)
1. Red Cloud, 1822-1909. 2. Oglala Indians—Kings and rulers
—Biography. 3. Oglala Indians—Wars. 4. Oglala Indians—
Government relations. I. Title.
E99.O3M347 2015
978.0049752440092—dc23
[B]

2015012736

Printed in the United States of America

The paper in this book meets the guidelines for permanence
and durability of the Committee on Production Guidelines for
Book Longevity of the Council on Library Resources.

Please visit our website at sdhspress.com

19 18 17 16 15   1 2 3 4 5

Cover and frontispiece: Red Cloud in 1880. Photograph by
Charles M. Bell. *Library of Congress*

Designed and set in Arnhem type by Rich Hendel

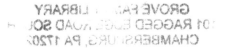

TO ANNE, LIZ, AND TOM

Contents

Acknowledgements

Especially helpful in the preparation of this book were notes amassed by my friends Thomas Powers in researching his recent book on Crazy Horse and R. Eli Paul when editing Red Cloud's autobiography. Eli also read the manuscript, as did Sioux experts Kingsley M. Bray and Ephriam D. Dickson III. Research in manuscript and newspaper collections in the Library of Congress and in military and Indian Office records in the National Archives complemented printed sources. In this regard, I wish to express my gratitude to the National Archives staff, especially retired employee Michael P. Musick and his colleagues: Michael Meier, Michael Pilgrim, and Todd Butler, and to the Library of Congress, especially Marilyn Ibach. Other institutions and staffs providing assistance were the National Anthropological Archives of the Smithsonian Institution; the Department of the Interior Library, especially Maureen A. Booth; the U.S. Army Military History Institute, Carlisle, Pennsylvania, especially David Keough and Kathy Olson; the Nebraska State Historical Society, especially the late Thomas R. Buecker and James E. Potter; the University of Nebraska Archives and Special Collections; the South Dakota State Historical Society, especially Nancy Tystad Koupal, Stephen S. Witte, and Ken Stewart; former director Ramon Powers and the staff of the Kansas Historical Society; the Colorado Historical Society; the State Historical Society of Iowa; the Montana Historical Society, especially former employee Charles E. Rankin; the Ohio Historical Society; and the State Historical Society of Missouri.

I also wish to thank the staffs of Fort Laramie National Historic Site, especially Sandra Lowry and Steve Fullmer; Little Bighorn Battlefield National Monument, especially John Doerner; the Denver Public Library, Denver, Colorado, especially Lisa Backman; the Casper Community College Library, Casper, Wyoming, especially Kevin Anderson; the Wyoming State Archives, Cheyenne, Wyoming, especially Ann Nelson and Cindy Brown; the American Heritage Center, Laramie, Wyoming, especially Rick Ewig, and the Sheridan County Fulmer Library, Sheridan, Wyoming, especially Karen Woinowski and Marci Mock.

Other special collections used included those of the Brigham Young University Harold B. Lee Library, Provo, Utah; the Beinecke Library, Yale University, New Haven, Connecticut; the Highland County Public Library, Hillsboro, Ohio; the Natrona County Public Library, Casper, Wyoming; the Johnson County Public Library, Buffalo, Wyoming; the Pioneer Museum, Douglas, Wyoming; the Rawlins Public Library, Pierre, South Dakota; the Southwest Museum, Los Angeles, California; the Newberry Library, Chicago, Illinois; the St. Charles Historical Society, St. Charles, Missouri; and the Colorado State University Library, Fort Collins, Colorado. Special thanks go to Margot Liberty, Elmer ("Sonny") Reisch, and Robert C. Wilson, former staff at Fort Phil Kearny State Historic Site; to Richard Young of the Fort Caspar Museum; and to retired National Park Service historians Jerome A. Greene, Paul L. Hedren, and Douglas C. McChristian.

To these individuals and institutions, I am forever grateful.

Preface

Perhaps no other American Indian leader has played so significant a part in the history of South Dakota as Red Cloud did. It should come as no surprise, then, that the Oglala chief is the subject of several major biographical studies. One of the best of Red Cloud's early biographers was George E. Hyde (1882–1968), a self-taught scholar who lived in Omaha, Nebraska, and brought a quick mind and useful experience to his work. Having lost his hearing and much of his eyesight as a young man, he faced an uncertain future. Hyde took up writing and chose Indians as his subject because, as he wrote to a friend, "I had sense enough to know that a man to write must have first hand information, and there was no subject as suitable as Indians."[1] The 1898 Trans-Mississippi Exposition in Omaha included an "Indian Congress," and young Hyde became acquainted with a number of its participants. He visited the Indians' camps, earned their trust, and deepened his interest in their past. As a writer, Hyde added another dimension to the study of western history by including the testimony of Indian witnesses.

Hyde's three books on the western Sioux—*Red Cloud's Folk* (1937), *A Sioux Chronicle* (1956), and *Spotted Tail's Folk* (1961)—are major contributions to the history of the West.[2] *Red Cloud's Folk* covers the period 1660–1878, describing Lakota movements onto the Northern Plains with a focus upon the Oglalas' interactions with other tribes and pioneer settlers up to their ultimate confinement on reservations. *A Sioux Chronicle* treats Oglala history during the reservation years, culminating with the tragedy of Wounded Knee. Although *Spotted Tail's Folk* deals with the Brules, Red Cloud often worked in concert with Spotted Tail, and many young

Brules fought under the Oglala chief's leadership during the struggle on the Bozeman Trail. Hyde's three books form a strong foundation for this present study.

Another important work on Red Cloud and his place in history is *Red Cloud and the Sioux Problem* (1965) by James C. Olson, a professional historian and former superintendent of the Nebraska State Historical Society. Olson had manuscript collections, government documents, and other printed sources easily available to him. His intent was to "provide an objective account of Red Cloud's role in the life of his people during the years when they were making the transition from warriors to wards of the Government."[3] Olson's sources dictated that the study would be from Euro-Americans' point of view and relate more of what they thought of Red Cloud rather than what Red Cloud thought of them. Olson verified Hyde's research and viewed his own study "as a supplement to Hyde's work rather than as an effort to revise it."[4]

The next step in Red Cloud's biographical history was the publication of an account of the chief's own memories of his life before the Bozeman Trail wars of the 1860s. Samuel Deon, a trader and an old friend of Red Cloud, held a series of daily talks with the chief at the Pine Ridge post office in 1893. Deon listened to Red Cloud's stories of his early life and then passed them on to postmaster Charles W. Allen, who fashioned the accounts into a manuscript. A former newspaperman, Allen published excerpts of the manuscript in three issues of *The Hesperian*, an obscure and short-lived Black Hills literary magazine, in 1895–1896. Allen later entrusted the manuscript to Addison E. Sheldon, early director of the Nebraska State Historical Society. The Red Cloud manuscript, thought to be Sheldon's own work, languished among Sheldon's papers for years.[5] In the mid-1990s, R. Eli Paul rediscovered the document's true value, which he described in a 1994 article for *Montana, the Magazine of Western History*. In 1997, the Montana Historical Society Press published the manuscript, edited by Paul, as *Autobiography of Red Cloud, War Leader of the Oglalas*.[6]

The discovery of the Red Cloud manuscript's true nature facilitated Robert A. Larson's major study of the chief's life. Larson believed that "Red Cloud remained a rather remote and misunderstood figure" and "continued to project such a blurred image" despite Olson's "excellent" work.[7] Using the autobiography and other materials that had appeared in the intervening period, Larson produced the model study *Red Cloud: Warrior-Statesman of the Lakota Sioux* in 1997. In the end, Larson saw Red Cloud as "a man forced by circumstance to fight a rearguard action, who contested every policy of the federal government to alter the Treaty of Fort Laramie" and "resisted every governmental interpretation that would mean a loss of territory for his people."[8]

Several other recent works have contributed significantly to telling the Red Cloud saga. Catherine Price's *The Oglala People, 1841–1849: A Political History* (1996) deftly explains the Oglalas' cultural framework and decision-making structures. *Crazy Horse and Chief Red Cloud* (2005) by Oglala writer Ed McGaa includes valuable oral history. Kingsley M. Bray, an English writer, published *Crazy Horse: A Lakota Life* in 2006. Finally, Thomas Powers's tour de force *The Killing of Crazy Horse* (2010) reveals much about Red Cloud's relationship with his erstwhile young follower who later became a rival.[9]

The most recent Red Cloud biography, *The Heart of Everything That Is: The Untold Story of Red Cloud, an American Legend* by Bob Drury and Tom Clavin, appeared in November 2013. Published by the old and reputable firm of Simon and Schuster, the book was widely publicized and acclaimed but has little value for the scholar. Direct quotations are the only items that are footnoted in the text, which abounds with egregious errors of fact and supposition. Long passages in which Red Cloud supposedly shares his thoughts on various events are particularly disturbing. The most blatant misrepresentation is the authors' account of Jim Bridger's alleged thoughts and deeds at Fort Phil Kearny on 21 December 1866, the day of the Fetterman fight, when it is known that the scout spent the winter of 1866–1867 at Fort C. F. Smith.

My own interest in Red Cloud has resulted in two previous books: *Circle of Fire: The Indian War of 1865* (2003) and the two-volume *Red Cloud's War: The Bozeman Trail, 1866–1868* (2010). The first of these volumes tells the story of Red Cloud's rise to lead the Lakota resistance to white expansion on the Northern Plains, while the second covers his fight against the fortification and use of the Bozeman Trail, a struggle that ended with the army's withdrawal from the disputed region and the Treaty of Fort Laramie in 1868.[10] The present biography adds other layers to the story, especially concerning the chief's childhood and youth. The writings of Doane Robinson, an early director of the South Dakota State Historical Society, have previously been dismissed as composites based upon the author's wide reading on Red Cloud, but Robinson's 1932 article in a little-known magazine includes an interview with the old chief conducted in about 1903. The account Robinson received from Red Cloud is now added to the story as a credible reconstruction of the chief's upbringing.[11] The Ella Deloria Collection at the Dakota Indian Foundation in Chamberlain, South Dakota, contains an account of the training that Oglala boys underwent in preparation for their adult role as warriors.[12] Deloria's material gathered from Oglala elders, together with Robinson's interview of Red Cloud, fills in a gap concerning the chief's youth.

Newspaper and magazine accounts also yielded new information and insights into Red Cloud's life. Much of this material was not readily accessible to previous biographers, but new technology has made important regional newspapers such as the *Cheyenne Daily Leader* and the *Omaha Bee* available in searchable form. Researchers can now read interviews with Red Cloud and other Lakota leaders on the reservations, quotations from the chief's speeches, and journalists' narratives of his actions at meetings and conferences. Newspaper reports of Red Cloud's dealings with government officials reveal his strong-willed personality and the many methods the Oglala leader used to make friends and influence enemies. Another important source was the *Council Fire*, a Washington, D.C., magazine devoted almost exclusively to Indian affairs. Founded by Alfred B. Meacham in 1878 and edited by Thomas A. Bland from 1882 until its demise in 1889, the journal was published as the *Council Fire and Arbitrator* from 1881 to 1886. The *Council Fire* printed many letters in which Red Cloud discussed his problems with the Office of Indian Affairs.[13]

Finally, I have attempted to create a larger cultural context for Red Cloud's particular way of viewing the world and its challenges, using his speeches and the pathbreaking work of anthropologists such as Clark Wissler, Alice C. Fletcher, James R. Walker, J. Owen Dorsey, Warren K. Moorehead, William K. Powers, Joseph Epes Brown, and Raymond J. DeMallie, as well as the writings of other Indian leaders and spiritual advisers. I have tried to let Red Cloud speak for himself as much as possible, fully realizing that his spoken and written words were filtered through interpreters of various levels of competence and, in his later years, through a private secretary to whom he dictated his letters.

1 Red Cloud's World

To die, defending truth, is a noble art
That wins, in time, the stoic mind's applause;
But Red Cloud played a hero's great part—
He lived, to strive for his own people's Cause.
—Leslie Thompson Dykstra[1]

On the day of the winter solstice in 1866, as the sun began to set in the west causing the temperature to plummet below zero, Red Cloud looked down on the mile-long ridge that gradually descended from the high hills that blocked the view of the soldiers at Fort Phil Kearny from the valley below. Strung out along the narrow pathway were the bodies of Captain William J. Fetterman and the eighty men who had left the little post on Piney Creek a few hours earlier. Their grotesquely stiffened corpses marked the ebb of the United States Army's attempt to keep the Bozeman Trail open and the zenith of Red Cloud's influence in the Powder River Country of what was then Dakota Territory and is now Wyoming. The Oglala strongman had inspired a coalition of several Sioux bands, the Northern Cheyennes, and the Northern Arapahos to achieve what few Indian warriors had ever done before—meet the army in a pitched battle and win. Now, he was the most powerful leader on the Northern Plains. The eastern press proclaimed him greater than Tecumseh or Red Jacket; his fame spanned oceans.

But it had not always been that way. Red Cloud's rise to power was as remarkable as it was swift, depending on his courage, persistence, and charisma. Like many heroes of old, he had earned his way, first becoming a heralded warrior, second a band chief, and third a leader who fought to preserve his people's way of life against invaders from the east. Although victorious for a time, he would eventually recognize that armed resistance could not succeed in the long run. Later in life, he would seek to maintain as much of his people's autonomy and cultural identity as possible through such means as negotiation, passive resistance, and selective adaptation to white men's ways. While his efforts achieved but limited success, his determination to defend his people's interests remained undaunted until his last days.

Red Cloud had been born into the Sioux Nation, the populous and aggressive confederation whose western warriors eventually became the most feared fighters on the Northern Great Plains. He was a member of the Oglala tribe, a subdivision of the Tetons or Lakotas, one of the Seven Council Fires of the Sioux.[2] The Lakotas had seven branches, several of which became famous in the Indian wars of the 1860s and 1870s. Besides the Oglalas, there were the Brules, Hunkpapas, Minneconjous, Sans Arcs, Two Kettles, and Blackfeet. Stephen R. Riggs, an early missionary, reported that Oglala meant "She Scatters Her Own," originating from a quarrel between two women. One threw some flour in the face of the other, and the victim and her friends, who thereafter were "called the Oglala," separated themselves from the others.[3]

As a group, the Lakotas had attained a position of power gradually. According to oral history, they had moved in the seventeenth century from southern Wisconsin, northwestern Illinois, and northeastern Iowa into present-day Minnesota, where the first written references to them occur. The Ojibways on their east acquired firearms, pushing the Lakotas to move westward until they reached the Missouri River around 1760, where they eventually obtained modern weapons of their own. Next, they acquired horses from Indians living to the south, increasing their mobility and hunting success. Horses also became the center of Lakota political and social life. Stealing them from one's enemies became the quickest way for a male to gain prestige, and horses were important gifts required to win a mate.[4] Horses also permitted the Oglalas to expand their pursuit of game into other regions; in doing so, they encroached upon the territory of adjacent tribes, and wars ensued. Contributing to the instability was the erratic movement of the bison herds, shifting from one area to another, causing hunters to follow them into other tribes' territories. Of course, the buffalo were the source of everything for the Sioux. The bulk of their food supply came from these animals, as did the skins that covered their lodges. Buffalo robes served as blankets; sinews made bowstrings; and even dried chips were used as fuel and, when mixed with dogwood bark, smoked in stone pipes.[5]

In the latter decades of the eighteenth century, the Oglalas and some Brules, always the vanguards of the western Sioux, began moving west across the Missouri River to find more abundant game. By the beginning of the nineteenth century, these Lakota pioneers controlled the Black Hills.[6] By 1820, they commonly spent part of the summer hunting in the valley of the North Platte River in southeastern Wyoming and western Nebraska.[7] Strong and powerful, the Lakotas were a proud people. According to the medicine man Red Hawk (born ca. 1829), they were "superior to all others of mankind."[8]

Among this people, Red Cloud was born in May 1821 along Blue Creek, a tributary of the North Platte River, opposite the mouth of Ash Hollow in present-day Garden County, Nebraska. His father was Lone Man, a Brule who had married Walks As She Thinks, a sister of the Oglala leader Smoke (1774–1864). When Red Cloud was about four years old, his father died, apparently from alcoholism, and his mother and her sisters raised him. Smoke became his mentor and surrogate father. Smoke's sixty-four-year reign as the head of the Hunkpatila band of Oglalas from 1800 to 1864 was the longest of any chief in Lakota history. Thus, the boy had a powerful protector and a special place in the Oglala tribe. However, Red Cloud was not in line to become hereditary chief—that honor went to Man Afraid of His Horses, Smoke's first-born, who, together with his own son, would later challenge Red Cloud's leadership, especially during the reservation years.[9]

As a Lakota, Red Cloud came into a world of connections, in which all things were related to one another. The Great Mystery or Spirit (*Wakan Tanka*) dwelt in every object, whether of nature or of man's making. As the Lakota philosopher Luther Standing Bear (1868–1939) put it: "From Wakan Tanka there came a great unifying life force that flowed in and through all things—the flowers of the plains, blowing winds, rocks, trees, birds, animals—and was the same force that had been breathed into the first man. Thus all things were kindred and brought together by the same Great Mystery." For this reason, Standing Bear declared, "The Lakota could despise no creature, for all were of one blood, made by the same hand, and filled with the essence of the Great Mystery."[10]

At the same time, the universe could not be fully understood or controlled by human beings. Supernatural forces abounded in the world, and Lakota people sought them as allies when they could.[11] As Red Cloud himself explained: "As a child I was taught the Supernatural Powers (*Taku Wakan*) were powerful and could do strange things; that I should placate them and earn their favor; that they could help me or harm me; that they could be good friends or harmful enemies." He learned to "gain their favor by being kind to my people and brave before my enemies; by telling the truth and living straight; by fighting for my people and their hunting grounds."[12] Because supernatural power was in all living things, each in a special way, Lakota people called upon certain animals to share their attributes, and warriors carried totems in honor of their affinity—a badger's claw for taciturnity in battle, buffalo horns for fearlessness, the tail of a coyote for cunning, or an eagle feather for speed, sight, and fierceness.[13] Because of this belief, Red Cloud and his people had a close relationship with nature, its elements, and its varieties of flora and fauna. When whites tried to take them away from their lands, they threatened not only Lakota livelihood but Lakota essence as well.

The Oglalas were also a circular people. They believed in the flow of things, seeing it in the ripples emanating from a stone thrown in a lake, the change of seasons from hot to cold and back again, the curved horizon of the plains, and the progression of men from swaddle to shroud. The circle was their archetype, and the conical tipi their physical representation of wholeness and unity. The camp circle was another sign of this unity. Their pattern of living was to move over the land from one place to another in chase of the buffalo and to harvest fruits and other wild foods from spring through fall. They were nomads, prizing territory over parcel, vastness over personal plot. They were roamers who took from the land but did little to change it. In contrast, Euro-American settlers brought the concept of private ownership of property with them from the Old World. The idea that each man had the right to keep a piece of land and do with it whatever he desired was something foreign and irrelevant to many Plains Indians.[14] The physical manifestation of this Euro-American belief was the house with four walls, a symbol of security and immobility, meant to protect the few who occupied it and keep out the uninvited. Red Cloud put the contrast with Lakota thinking succinctly when he said, "Where the tipi was, there we stayed and no house imprisoned us. No one said, 'To this line is my land, to that is yours.'"[15]

Euro-Americans also sought to alter the land to suit their own purposes. George Armstrong Custer's brother-in-law, First Lieutenant James Calhoun, provided an example of this thinking. While in the Black Hills with Custer in 1874, Calhoun wrote: "It is a great pity that this rich country should remain in a wild state, uncultivated and uninhabited by civilized men. Here the wheel of industry could move to advantage. The propelling power of life in the shape of human labor is only wanting to make this a region of prosperity."[16]

Red Cloud did not share Calhoun's vision. Believing that "the supernatural powers, *Taku Wakan*, had given to the Lakotas the buffalo for food and clothing," he had no wish to "dig the earth to make food and clothing grow from it."[17] With the invaders from the east seemingly determined to destroy the Lakota way of life, it is little wonder that Red Cloud chose to resist them.

2 Preparation for Greatness

As a young man he was a terror in war with other tribes.
—James H. Cook[1]

From birth, Oglala boys began the learning process that would fit them for life. Survival required knowledge of the land, including its geography, its flora and fauna, and all the resources it offered. The Great Plains seemed barren to Euro-Americans at first sight, but Plains Indians had learned to use this environment to their advantage. They knew how to subsist on indigenous foodstuffs, knew where to locate water in the dry season, and knew the best places to hide or defend themselves from adversaries. During the first decade of the twentieth century, Red Cloud told South Dakota historian Doane Robinson about his childhood training and activities. Robinson published his Red Cloud interview in a magazine article in 1932. Because it may be the only account of the chief's childhood in something approximating his own words, it is quoted extensively here:

> When I was still hanging in the cradle, my mother . . .
> sang to me until I slept; and when I was awake and
> the birds sang, my mother would whistle just like the
> bird, and so before I could talk I learned to imitate the
> birds. . . . When I was two I knew all the birds by their
> songs and by their plumage and I have not forgotten.
> I knew why my father wore an eagle feather in his war
> bonnet and that year I plucked it from his bonnet to
> decorate my dog. So much my mother taught me about
> birds and animals, flowers and plants. I was still very
> young when I knew all the plants that were good for
> food and all that were good for medicine. . . .
>   There were many games which we were encouraged
> to play, for in that way we developed our muscles and

trained them to act as we chose. Two games in particular were the pastime of the boys. The first was played with spinning tops and the winner was the one who kept his top going for the longest time and had it fall nearest some given object when it stopped. The little boys became very expert in this. As they grew older they were given the game of javelins . . . and the boys became very expert in throwing javelins, or spears, accurately and for a long distance. . . .

We traveled a great deal to visit distant relatives, for hunting and to obtain food; to council with other bands or for pleasure, so that as a young boy I knew much of the country from the Republican on the south to the Yellowstone on the north; from the Missouri on the east to the mountains on the west, and so I learned about the birds, animals, and plants in all this extensive region. Our life depended upon knowledge of this kind and upon ways to make our escape from any place, in time of war or trouble. When I grew older I was sent alone on long trips to test the ability to find myself wherever I might be.[2]

While Red Cloud's sharing of his own memories with Doane Robinson provides a unique window into his childhood experiences, accounts of others can also shed some light upon them. For example, according to the Sioux author Charles Eastman, Red Cloud exhibited great promise in his youth by showing patience and resolution, two much-prized values among his people. His ability with horses was a case in point. At six years of age he was given a colt and encouraged to sit on its back "without saddle or bridle," for a boy who could master a wild animal would, as an adult, "be able to win and rule men."[3] He gentled his colt without help and reportedly had a special affinity for horses for the rest of his life. Witnesses often remarked that Red Cloud's horses were well broken, a testament to his patience with the animals. Eastman's telling of the story of the twelve-year-old Red Cloud's first attempt to kill a buffalo shows the future chief's resolve. When one of his arrows did not penetrate his target more than a few inches, he dashed up to the huge animal and tried to push the projectile in farther.[4]

Like other Oglala boys, Red Cloud received warrior training. While Red Cloud's interview with Doane Robinson included few details of this training, other sources describe common experiences that an aspiring Oglala warrior was likely to have. The ideal of the warrior, Oglala boys were told, was summarized in the saying, "It is better to die on the battlefield than to live to be old."[5] Bravery was the greatest of the virtues to which a young warrior should aspire. From the beginning of a boy's life, he heard his elders tell stories of victory in battle. He would be given a small bow and arrows, which enabled him to kill birds and small animals to improve his marksmanship.[6]

When the men of the village decided a boy was old enough, they put him through a number of tests, such as carrying water, finding animals, or scouting to discover something of importance. These trials were meant to increase his resourcefulness and make him watchful. Adult males instructed boys how to earn recognition and honor in war by stealing the enemy's horses, counting coup (physically touching an enemy or a dangerous place), or by bringing wounded or dead tribesmen off the field of battle. To be recognized as a respected warrior, a man needed to carry out such acts of bravery multiple times, preferably in the sight of witnesses. Lakota society developed an elaborate system of war honors, and young men were strongly encouraged to prove their worth in battle. Boys accompanying a war party were often assigned the task of carrying water for the group, a job suited for their capabilities, which also permitted them to learn by example.[7]

Oglala Lakota boys were also instructed in their people's expectations for leaders. War honors were a path to respectability in society but did not suffice to make a chief. Other virtues, such as generosity, hospitality, and fortitude, were expected of a leader as well. A man should demonstrate his generosity and hospitality by providing feasts for the community. Once he had done so enough times to establish his reputation, his opinions would carry weight in council. Once recognized as a chief, he would be expected to continue demonstrating his generosity by organizing feasts for the old men of the tribe, "for no stated reason," as the twentieth-century Sioux scholar Ella Deloria put it. A chief who did so would retain great prestige into old age, even "though he may be poor and helpless." According to Deloria, "this was the old teaching to the youth of the Teton Dakota [Lakotas]," and as such, was the challenge to succeed.[8]

Red Cloud, by both his own account and those of other witnesses, worked hard to achieve the Oglala standard of success in war. He reportedly joined his first war party at age fourteen and took his first scalp at fifteen or sixteen. He was part of an expedition that killed some eighty Pawnee Indians about 1837, according to information obtained by photographer and ethnologist Edward S. Curtis several decades later. Curtis's sources credited Red Cloud with taking two scalps in that campaign. Two years later, Red Cloud led a war party of his own that inflicted eight additional casualties. Red Cloud reportedly killed two Shoshoni and ten Crow Indians in other battles.[9] James H. Cook, a friend of the chief in his later years, recorded that "old warriors who fought by his side have told me of his killing five Pawnees in one fight, using only his knife as a weapon after sending one arrow from his bow."[10] Cook also wrote that Red Cloud told him the story of how he dispatched five Crow enemies in one day, each coming to meet him in single combat. According to information gathered by Indian agent Valentine T. McGillycuddy and Charles W. Allen, Red Cloud could claim eighty coups in his career. Edmund B. Tuttle, who accompanied Red Cloud to Washington in 1870, wrote that the chief claimed to have fought in eighty-seven battles, and had often been wounded.[11] Red Cloud's most serious injury, according to Cook, was due to a Pawnee arrow that completely penetrated his body, with the "head protrud[ing] a couple of inches near his backbone." Though unconscious for three full days, he eventually recovered from the wound.[12]

It is clear from the historical record that Red Cloud not only achieved great renown among his people for his prowess as a warrior but also took pride in his achievements in battle. Thisba Hutson Morgan, who taught school in 1890–1891 at the Pine Ridge Agency where Red Cloud then lived, wrote of meeting the old chief whose "most prized possession" was a buckskin war shirt "fringed with the long scalp locks of Crow Indians."[13] In 1931, James Cook collected some of Red Cloud's war songs that had been passed down to his descendants. Translated into English, one song reads, "The coyotes howl over me. That is what I have been hearing. And the owls hoot over me. That is what I have been hearing. What am I looking for? My enemies. I am not afraid."[14]

Red Cloud did not always triumph as a warrior; he knew reverses as well. Charles Eastman once asked Red Cloud if he had ever been afraid. The chief told Eastman about his experience as a sixteen-year-old on a buffalo hunt "in the Big Horn country [possibly northeastern Wyoming]," where the hunters had to be on their guard against Crow or Shoshoni enemies. Red Cloud pursued and killed a buffalo bull on his own and worked alone to cut up the carcass and pack the meat for transport to the camp while remaining vigilant in case of attack. Suddenly, as Eastman relates the story, Red Cloud "heard a tremendous war whoop, and glancing sidewise, thought he beheld the charge of an overwhelming number of warriors." The young warrior tried to give an answering war cry, but "instead a yell of terror burst from his lips, his legs gave way under him, and he fell in a heap." He was mortified to discover that the enemy war shout had been only "the sudden loud whinnying of his own horse" and that the warriors had been "a band of fleeing elk." Red Cloud was so ashamed of this experience that he never spoke of it until his talk with Eastman many years later.[15]

When Red Cloud was about fourteen, his people's circumstances changed in an important way. In 1834, fur traders William Sublette and Robert Campbell established Fort William, better known as Fort Laramie, a log-stockaded trading post near the confluence of the Laramie River and the North Platte River a few miles west of the present Wyoming-Nebraska border. The post, whose physical layout evolved over several decades, became a crucial rest and supply stop on the Oregon, California, and Mormon emigrant trails in the 1840s. While building the post, Sublette and Campbell sent messengers to the Oglalas encouraging them to trade there. In 1835, Oglala bands set up camp near the fort, which became an important center of tribal activity for the next forty years. The post contained much that the Lakotas desired—firearms, gunpowder, metal goods, brightly colored blankets, beads, and the like. Although the military took over the fort in 1849, it continued its role as a material oasis for the Indians of the region.[16]

At Fort Laramie, Red Cloud took advantage of opportunities to observe white people. Mary Gardiner, his sister's daughter, had married John Baptiste Richard, Sr., an influential trader in the region in about 1843. John's younger brother Peter married the daughter of an Oglala headman. Consequently, Red Cloud's family was intermingled with white traders.[17] Because Smoke's band often made its home around Fort Laramie, Red Cloud also came to know another kind of white man, the United States soldier, whom he would later meet in deadly combat. As Red Cloud later explained to Doane Robinson, "It was the Great Father's [the president's] soldiers that came about that time that interested me most. I stayed as near to them as I could and learned much of the way in which the soldiers were managed and the discipline required of them." The young warrior immediately saw the differences between the Lakota and white approaches to warfare. The Oglalas trained their warriors to be self-reliant. Their fighters never stood in linear ranks "to be shot down by the enemy," as Red Cloud put it. While Red Cloud learned much that was useful in his later battles with the United States Army, he told Robinson "the thing that was most useful was my knowledge of what the soldiers were trained to do." He saw no need to adopt the whites' methods himself.[18]

As the Lakotas began to explore north of Fort Laramie, they penetrated Crow country, and, in the late 1830s and 1840s, the Powder River became the approximate boundary between the Oglalas and their long-time enemies. This region would become a battleground with both sides contesting for its resources, and it was there that Red Cloud achieved his martial reputation among his people.[19] As his friend James Cook later put it, Red Cloud "became a terror in war with other tribes."[20]

It was during this period of intense combat that Red Cloud obtained his permanent name. There are conflicting versions of its origin and meaning. Indian agent Valentine T. McGillycuddy stated that, according to Indian informants, the Oglala leader "took his name from the claim [that] when on the war path with his warriors their red blankets covered the hillside like a red cloud," while historian George E. Hyde believed that the name came from a meteorite that blazed across the sky on the day of Red Cloud's birth.[21] According to Charles Allen's manuscript based on Red Cloud's conversations with trader Sam Deon, the name had been used by the chief's father and grandfather in keeping with the custom among the Sioux of using multiple names during a lifetime. When Red Cloud was born, an older cousin had the name, but in 1837, that warrior died in a fight with the Pawnees. When the sixteen-year-old joined a war party to avenge the death of his cousin, someone shouted, "Red Cloud's son is coming. Red Cloud comes," and the name was his.[22] The youth's name prior to that time has been reported variously as Tall Hollow Horn and Two Arrows.[23]

The training and experiences of his early years prepared Red Cloud for his leadership role to come. He was imbued with the Sioux ethic of social responsibility, tempered by the knowledge of his own vulnerability, and possessed an unwavering loyalty to the welfare of his people. Red Cloud came of age during a period of dynamism. His early life coincided with the time when Oglala warriors rode free in the chase for buffalo and for battle honors. Few have ever had this freedom, and the warriors of the Plains left a legacy that still reverberates in word and song.

3 Rise to Power

*A Ute Indian was crossing a stream on a wounded
horse and the horse gave out and the Ute was drowning
. . . when Red Cloud rode in on his sprightly horse and
grabbed the Indian by his hair and brought him to
shore. When he arrived there holding on to his hair
he took his knife and slashed off his scalp and let the
Indian fall to the ground and he rode off with the scalp.*
—American Horse[1]

For Red Cloud, the years between 1841 and 1855
were a time of great achievement. The young warrior grew
into a man of prominence, well on his way to becoming a
chief. In doing so, he lived up to many of the aspirations his
boyhood training had inculcated in him. He did not escape
sorrow, however, as a great tragedy accompanied his rise to
power.

Red Cloud grew up in a world of intrigue and violence.
Not only did the Oglalas fight the Pawnees, Omahas, Crows,
Utes, Shoshonis, and other non-Sioux tribes, they experienced rivalries, disputes, and divisions among themselves.
By the mid-1830s, Bull Bear had emerged as a leader of a
faction later called the Cut-Offs, while Smoke's people were
known as the Bad Faces.[2] Red Cloud achieved notoriety
among the Oglalas in 1841, when at the age of twenty, he
killed Bull Bear, then regarded as the most powerful Oglala
chief—and a bitter rival of his uncle, Smoke. The animosity
between the two chiefs was reportedly exacerbated by the
presence of liquor. The killing occurred when Bull Bear's
band, then known as the Koyas, was camped near Smoke's
Bad Faces in the vicinity of Chugwater Creek, about forty
miles south of Fort Laramie.[3] Red Cloud's account has it
that a young Bad Face warrior, who was not liked by Bull
Bear and other Koyas, stole a Koya woman. This theft was

an acceptable way of taking a wife if the bride's family decided to approve the match. They did not, however, and regarded the stealing as an insult. Bull Bear and some followers, under the influence of liquor, went to the Bad Face village and killed the young groom's father. Red Cloud and his fellow Bad Faces retaliated, and Red Cloud dispatched the wounded Bull Bear with a shot to the head.[4]

Red Cloud's version of these events differs markedly from the accounts of Francis Parkman and Rufus B. Sage, who record that Bull Bear was killed trying to end a drunken brawl while visiting Smoke's camp. Red Cloud's story also leaves out a possible motive for the killing. According to George E. Hyde, Bull Bear had been enraged when white fur traders encouraged Smoke to oppose his authority and had challenged his rival to fight. When Smoke declined to come out of his lodge, Bull Bear allegedly killed Smoke's favorite horse. Smoke's friends and family, "young Red Cloud among them," reportedly plotted revenge.[5]

Whatever Red Cloud's motive or role, the killing of Bull Bear gave him more prominence than a twenty-year-old Oglala warrior would usually have. He built up his reputation further with battle exploits, placing himself on the path to success. Red Cloud apparently led a war party against the Pawnees not long after the killing of his uncle's nemesis. After Bull Bear's death, the Koya band apparently adopted a new name, Kiyuksa, known in English with the perhaps-inaccurate translation "Cut-Offs."[6]

Another result of the killing was that Man Afraid of His Horses, who had assumed the functional role as the successor to Smoke in Oglala society, consulted Red Cloud in all important matters. Born in about 1802, Man Afraid of His Horses became head of all the Oglalas following the death of Conquering Bear, the government-recognized chief of all Lakotas, who had been mortally wounded in the 1854 Grattan Fight. The Oglala chief's name came from his reputation as a warrior so feared that the enemy trembled at even the sight of his horses. Whites garbled the translation to mean the opposite. When Smoke died in 1864, Man Afraid took over as leader of his band, now known as the Smoke People. Bull Bear's son, Little Wound, eventually became chief of the Cut-Offs, and though the groups sometimes worked together, Little Wound bitterly resented Red Cloud and set an independent course when he could afford to do so.[7]

Sometime between 1848 and 1851, as his battle honors accumulated and his stature among his peers grew, Red Cloud sought to marry. His decision was entirely consistent with Lakota social norms for a man in his twenties.[8] In the Sioux social order, marriage "became as much an alliance between families as the joining of two individuals."[9] The male suitor offered property to his intended bride's parents, usually horses because of their great value in a nomadic, buffalo-hunting society. Red Cloud's exploits had won not only admiration from his fellow warriors but also the favorable attention of many young women. Two, in particular, captured Red Cloud's fancy—Pretty Owl (sometimes called Painted Owl) and Pine Leaf—and they returned his interest. He finally decided to try for them both, which was allowable among the Sioux due to the paucity of male mates in a warring society. However, social convention prohibited marrying both women at once; each required a separate ceremony, with at least a month or two in between. Red Cloud decided to marry Pretty Owl first, and, after some negotiation, the young man provided enough horses to win his bride. He did not inform Pine Leaf of his intention to make her his second wife later. The marriage to Pretty Owl took place when the Oglala village was located at the base of Raw Hide Buttes, near present-day Lusk, Wyoming.[10]

When Red Cloud left his lodge the morning after his wedding, a horrific sight confronted him. Pine Leaf was hanging from a tree branch, having taken her own life in despair. Pine Leaf's relatives destroyed Red Cloud's lodge and Red Cloud went into deep mourning. Pine Leaf's death and its aftermath caused Red Cloud to resolve never to have more than one wife, and apparently he never did.[11] His friend and advisor, Nick Janis, reported that another woman did arrive in his life much later. As he put it, "One come once, but she didn't stay," indicating that Pretty Owl had driven her off.[12]

The daughter of Hollow Bear and Good Owl, Pretty Owl was also known as Mary Good Road in later life. Apparently, the union of Pretty Owl and Red Cloud was a strong, enduring bond. As Nick Janis described it about 1891: "He don't stand out for nobody, 'cept, perhaps, his squaw. She makes it lively for him sometimes."[13] Underlying Janis's characterization was that Sioux society valued women in a variety of ways poorly understood by white observers then and now. They were life-givers, rearing their children according to four key virtues: fortitude, generosity, bravery, and wisdom. Women were the food gatherers and clothing makers. In the traditional matrilineal society of the Oglalas, women owned the tipis and camp equipage. Lakota women also played essential roles in religious ceremonies as seers and as healers. White observers in the nineteenth century generally did not recognize the importance of these roles, instead seeing Oglala women as drudges, burdened and oppressed by constant labor.[14]

Red Cloud and Pretty Owl may have had as many as ten children. Six, a son and five daughters, were known to be living at the time of their father's death in 1909. A probate document listed their names as Jack Red Cloud, War Bonnet, Leading Woman, Plenty Horses, Charges at Him, and Tells Him. Jack Red Cloud succeeded his father as chief of the Bad Faces but did not achieve the same influence as his father in tribal affairs. Another daughter, Comes Back, appears in the 1887 Pine Ridge Agency census. She was twenty-two years old and listed as residing with her parents. Her absence from Red Cloud's probate record may suggest that she died before 1909. The 1887 census also listed a nine-year-old son of Red Cloud named Comes Growling, who, like Comes Back, is not named among the chief's heirs in the 1909 probate record. In 1891, George Bird Grinnell recorded that another son of Red Cloud had died in about 1880 and was buried in a mission cemetery at Pine Ridge. In 1892, a Pittsburgh newspaper reported that a son of Red Cloud, named Big Cloud, was sent home from Hampton Institute in Virginia because he had an advanced case of tuberculosis. Thus, Pretty Owl's 1904 statement to Indian agent John R. Brennan claiming one son and five daughters may have referred only to living descendants.[15]

The precise timing of Red Cloud's emergence as a full-fledged chief has been somewhat murky. He completed an important prerequisite to be recognized as a chief by his people, namely, the *Hunka* (or *Huka*) ceremony, about 1855. The event required Red Cloud to organize and provide for a large feast and to give away practically all his wealth. It conferred great prestige but did not make him a chief in and of itself.[16] It was apparently Red Cloud's refusal to accede to the United States government's desire to keep the Bozeman Trail open in 1866, combined with his military successes in the war that followed, that finally cemented his status as a chief recognized by Lakotas and whites alike.

4 Road to War 1850–1865

The white man's steps are strong, and have crushed us as the grass is crushed by the feet of the buffalo.
— Red Cloud[1]

At mid-century, Red Cloud and his people began to feel the impact of white migration. In the mid-1840s, covered-wagon emigrants passed through present-day Nebraska and Wyoming bound for Oregon, California, and Utah, creating new pressures on Northern Plains Indians for subsistence. In 1843, the first year of significant emigrant traffic, only a thousand or so pioneers used the Platte River Valley as a road to points further west, but the numbers grew to nearly seventy thousand in 1852, partially as a result of the 1848 discovery of gold in California. As the total continued to rise, the Lakotas' initial friendly attitude toward whites changed. Fur traders, the first whites to arrive in Lakota country, had brought firearms and other material goods that benefited Lakotas. The white travelers who came in increasing numbers, however, scared away or hunted the game, burned the wood along the river banks, depleted grasses with their livestock, thus depriving Indian ponies and bison alike of grazing areas, and, in 1849, brought deadly diseases with them such as cholera, measles, and smallpox.[2] White presence in Lakota country gradually became more of a burden and less of a blessing. Red Cloud later explained the Lakota perspective with characteristic directness when, during his first visit to Washington in 1870, he addressed President Ulysses S. Grant: "Suppose I should go to your country, tear down your fences, and steal your cattle and your hogs, would you stand by and have no word to say? No, Father, I know you would not. In all the troubles of my people, the white man has been the first aggressor."[3]

Because many tribes expressed such concerns about white emigration in the 1840s, government officials felt the need to protect white emigrants from conflict that might result from rising Plains Indian resentment. To this end, the federal government established three forts along the overland migration route—Fort Kearny (1848) in present-day Nebraska, Fort Laramie (1849) in modern Wyoming, and Fort Hall (1849) in what is now Idaho. In September 1851, representatives of the United States government met with Sioux, Cheyenne, Arapaho, Shoshoni, and Crow councils at Horse Creek, near Fort Laramie, to seek a formal agreement concerning whites' travel through the area. In the end, the Indians agreed to allow emigrants safe passage and acknowledged permanent military posts along the migration route in return for annual payments of fifty thousand dollars in trade goods. The parties signed the document on 17 September 1851.[4]

From the government's perspective, the Horse Creek agreement quieted conflicts by defining general areas each tribe claimed and asking each to respect the others' lands. There was one problem, however. Many Sioux coveted the area described in the treaty as Crow country because they looked to the north and west for new hunting grounds free from white pressure. The 1851 agreement set aside swaths of present-day southeastern and south-central Montana, as well as much of northern Wyoming, for Crow use. This vast territory, covering about thirty-eight million acres, was abundant in natural resources. Fur trader Edwin Denig called the portion east of the Bighorn Mountains "perhaps the best game country in the world."[5] There one could find succulent grasses, wild fruit, streams teeming with fish, and buffalo, antelope, and deer in large numbers. Bears roamed the region, while mountain lions, beavers, and rabbits added further variety. Bighorn sheep were present in the mountains that took their name. The Crow lands were truly a great cornucopia for tribes hard-pressed by white encroachment. The Oglala people were well aware of these advantages.[6]

White travelers' impact on the Oregon Trail caused serious trouble for the western Sioux, and, as a result, on 15 June 1853 tensions came to a head when Minneconjou Lakota warriors seized the ferry emigrants used to cross the North Platte River near Fort Laramie. An army sergeant retrieved the boat but came under fire in doing so. Soldiers entered the Indian camp, seeking to arrest the perpetrators. A fight ensued in which the troops killed several warriors and took two prisoners.[7]

An even more serious confrontation took place on 19 August 1854, in which brash young Second Lieutenant John L. Grattan, an inebriated interpreter, and twenty-nine Fort Laramie enlisted men died. A dispute over the death of a cow belonging to Mormon emigrants was the catalyst, but the tragedy resulted from army officers' arrogance and ineptitude, combined with the drunken interpreter's needless provocation of the Sioux. According to fur trader Frank Salaway, the Brule leader Conquering Bear attempted to defuse the situation by allowing the emigrant whose cow had allegedly been stolen by a Minneconjou visitor to his camp to choose any horse he liked from the chief's own herd as compensation. Lieutenant Grattan, a recent West Point graduate with little experience in negotiating with Indians, rejected Conquering Bear's proposal and demanded the Minneconjou's surrender. When his demands were not met to his satisfaction, Grattan opened fire on the Sioux camp. The fight that followed resulted in the deaths of Grattan and all his men. By his own account, Red Cloud did what he could to prevent the conflict, but after Grattan's soldiers fired at the Sioux camp, the young warrior and his comrades joined in the fighting. Conquering Bear, who had been recognized by the government as paramount chief of all Sioux at the Horse Creek council of 1851, was mortally wounded.[8]

On 3 September 1855, in retaliation for the Grattan disaster, Brevet Brigadier General William S. Harney's forces attacked Little Thunder's band of Brules on Blue Water Creek in present-day Garden County, Nebraska, killing eighty-six people and capturing seventy women and children. Although Red Cloud was not present at this clash, he surely knew of it. Following the attack at Blue Water Creek, Harney summoned Lakota and Yanktonais leaders to a council, which met at Fort Pierre in present-day South Dakota in May 1856. The general attempted to impose a treaty on the tribes that would guarantee no further Sioux presence on the Oregon Trail. Harney also told each tribe to appoint a head chief to deal with the government in the future.[9]

While Harney may have reasoned that his actions would subjugate the Lakotas, they did the opposite. A large Lakota council assembled near Bear Butte in the Black Hills in 1857 to consider an appropriate response. The consensus was to exclude whites, other than traders, from the region north of the North Platte River and west of the Missouri. No further treaties would be made, and the Lakotas would go to war with the Crows to gain control of the buffalo country near the Powder River.[10]

Red Cloud's personal role in the decisions made at Bear Butte is largely a matter of conjecture. While he was clearly a respected man among the Oglala Lakotas due to his distinguished war record, his voice would be only one of many influential voices in a large council. Further, Red Cloud had not yet achieved a status equal to that of a hereditary chief in the 1850s. Even so, little doubt exists that he was involved in carrying out the council's wishes. Red Cloud had been among the emissaries that circulated the summons to council at Bear Butte, and he led one of the first war parties against the Crows after the 1857 council. In 1893, Red Cloud told his friend Sam Deon about some of his battles against the Crows dating to the 1850s and 1860s, including one in which he lost a nephew.[11]

With the aid of Cheyenne and Arapaho allies, the La-kota war against the Crows achieved significant successes. By 1860, the allied forces had gained complete control of the Powder River region.[12] Margaret Carrington, the wife of an army colonel, recorded that when military authorities asked the Cheyennes in 1866 why they and the Sioux had gone to war against the Crows, Cheyenne chiefs answered: "The Sioux helped us. We stole the hunting grounds of the Crows because they were the best. The white man is along the great waters [the Missouri], and we wanted more room. We fight the Crows because they will not take half and give us peace with the other half."[13] Simply put, the land was a much-desired prize as a haven from white pressure, and it was worth fighting for.

The Lakotas and their Cheyenne and Arapaho allies would not be able to enjoy the fruits of their conquests for long. American prospectors made gold and silver strikes in present-day Colorado in 1858, followed by additional discoveries in what would become Montana and Idaho over the next five years. These regions became magnets drawing fortune-seeking whites in large numbers, some of whom wished to cross the Lakotas' new sanctuary en-route. Unsurprisingly, the United States government soon considered seeking new roads across Sioux country once more, dispatching an expedition under Captain William F. Raynolds in 1860 with orders "to mark out a wagon route connecting the Oregon Trail and the Yellowstone-Missouri Basin."[14] The Raynolds expedition route approximated that of the Bozeman Trail. In 1864, John M. Bozeman and John M. Jacobs led the first wagon train to Montana on this road, leaving the Oregon Trail near present-day Casper, Wyo-ming, and proceeding north along the east side of the Big-horn Mountains. After crossing the Powder River, the route angled northwest to the Yellowstone River and then west to the gold fields. Because it threatened his people's hunting grounds, Red Cloud would fight to close this route.[15]

Adding to the Lakota people's problems was the struggle of their eastern kin in Minnesota and the newly organized Dakota Territory. The United States-Dakota War of 1862 resulted in a catastrophic defeat for the eastern Sioux and the loss of the bulk of their lands in Minnesota. Some eastern Sioux sought refuge among the western tribes. The news these refugees carried was not conducive to friendly Lakota-United States relations.[16]

Realizing that whites now coveted the last sanctuary of his people in the Northern Plains, Red Cloud decided to go to war against the whites. The following speech apparently delivered in 1864, explains his thinking: "My countrymen, shall the glittering trinkets of this rich man, his deceitful drink that overcomes the mind, shall these things tempt you to give up your homes, or hunting grounds, and the honorable teachings of our old men? Shall we permit ourselves to be driven to and fro—to be herded like the cattle of the white men?"[17]

According to George Sword, a council agreed with Red Cloud, and a crier went through the village proclaiming that from that time on the people would fight the white men and that warriors who did so would receive the same honors as they would fighting rival tribes. In this manner, the Sioux symbolically transformed whites into enemies. From then on, they were not to be respected but killed. Sioux oral history records that as fighting along the Bozeman Trail progressed, Crazy Horse told the young men that they could no longer afford the luxury of counting coup. They must fight to kill whites and kill them quickly.[18]

The Bozeman Trail was far from the only concern for the Sioux, however. On 29 November 1864, Colorado volunteer cavalry troops under Colonel John M. Chivington attacked a peaceful Southern Cheyenne and Arapaho village located on Sand Creek in present-day Kiowa County, Colorado. Recruited to fight Indians, most of Chivington's troops were near the end of their hundred-day enlistments and lacked the discipline of regular soldiers. The result was catastrophic. Chivington's dawn raid completely surprised the village. The colonel had ordered his men to take no prisoners, and the carnage that followed shocked even some of the most hardened Indian-haters. When fighting ceased, 53 men and 110 women were dead. The Colorado volunteers later displayed scalps and severed genitals to cheering crowds in Denver. When news of the slaughter reached the East, several investigations followed. Chivington escaped military prosecution by resigning his commission. After the initial shock dissipated, the survivors sought revenge.[19]

Following the Sand Creek Massacre, envoys representing the survivors went north to the headwaters of the Powder River to inform the Oglala, Minneconjou, and Sans Arcs Sioux, Northern Cheyennes, and Arapahos of the disaster. The messengers asked these northern bands to participate in a campaign of reprisal. Northern and Southern Cheyennes, separated for several decades by circumstances of geography and trade, came together for one last time. The western Sioux, beginning to feel the pressure of white encroachment in the north, joined their relatives and allies, bringing large numbers and fierce fighters to the fray.[20] During the extended conflict that followed, Red Cloud emerged as the most influential Indian on the Northern Plains.

As news of Sand Creek traveled to northern tribes, Indian groups closer to the massacre site chose Julesburg, Colorado, located on the South Platte River just south of the present Colorado-Nebraska border, as their first target for a retaliatory strike. Attacking on 7 January 1865, a force of Cheyennes, Arapahos, and Lakotas killed fifteen soldiers of the Seventh Iowa Volunteer Cavalry Regiment stationed at nearby Fort Rankin. Returning on 2 February, the raiders torched the town, keeping the Iowa volunteers inside their adobe-walled garrison.[21]

Receiving word of the attack, Ohio volunteer troops rode east from Fort Laramie. The soldiers found the war party in western Nebraska and fought two skirmishes in early February against the large body of warriors moving north and west to join their allies in the Powder River Country. Casualties were few on both sides, but the overwhelming numbers of Sioux and Cheyennes made it foolhardy and futile for the soldiers to continue their pursuit.[22]

Skirting the east side of the Black Hills, the warriors reached Red Cloud's camp on the Powder River and enlisted his aid. The timing was propitious, as Red Cloud had already decided that he must fight the white invaders at every opportunity. Together the allies decided to move south and attack the military post known as Platte Bridge Station, which guarded a crossing of the North Platte River near present-day Casper, Wyoming. After preparations, the warriors began the journey. George Bent, the son of an American fur trader and a Cheyenne woman who fought alongside his mother's people, later estimated that the force numbered three thousand warriors and formed a line stretching two miles across the plains. Oglalas, Minneconjous, Brules, Sans Arcs, Cheyennes, and Arapahos were included in the group. Red Cloud and Man Afraid of His Horses led the Oglala contingent, and the young Crazy Horse was also among them.[23]

The combined Indian forces arrived in the vicinity of Platte Bridge Station in late July 1865. On 25 July, the chiefs selected ten men to act as decoys, hoping to draw the garrison into an ambush. The stratagem did not succeed, and there was only intermittent fighting that day with few losses on either side. The next day, however, the Indians wiped out a detachment of soldiers guarding wagons returning from a supply run to an outlying post. This engagement took place at a spot known as Red Buttes, a few miles west of Platte Bridge Station. A cavalry detachment from the station under Lieutenant Caspar Collins attempted to rescue the supply train, but the Indians drove his force back to the post. Total army casualties for the day included Lieutenant

Collins and twenty-seven men killed and nine or ten men wounded. The army later memorialized Collins by changing the name of Platte Bridge Station to Fort Caspar, which became the namesake for the modern city of Casper, Wyoming. The fighting on 26 July represented one of the worst defeats that volunteer troops suffered at the hands of the Plains Indians during the Civil War period. On 27 July, the Indians again showed themselves, destroyed five miles of telegraph lines, and left.[24]

It would not be accurate to describe Red Cloud's role in the Platte Bridge Station fights as that of a "commander" in the sense commonly understood in the United States military. Plains Indian war-party leaders, such as Red Cloud in this instance, were responsible for leading their men to the battlefield, but the specific manner in which the warriors fought once the battle began was generally left to each man's discretion, as each sought his own battle honors. War-party leaders would arrange ambushes for their adversaries as a matter of course, but they could exercise comparatively little control over their warriors once the trap was sprung.[25]

In late July, troops under the command of Brigadier General Patrick E. Connor started north from Fort Laramie and from Omaha, Nebraska, in an effort to punish Sioux, Cheyenne, and Arapaho Indians for the sack of Julesburg and for other raids they had carried out in retaliation for the Sand Creek Massacre. Connor's force, known as the Powder River Expedition, totaled about two thousand men. Most were Civil War volunteer troops, not regulars, and many wanted their discharges badly. As they saw it, the war they had volunteered for was over. Connor's forces were divided into three columns, which were to approach the Powder River Country from different directions and rendezvous on about 1 September in the vicinity of the Tongue River, a stream that rises in the Bighorn Mountains of northern Wyoming and flows into the Yellowstone near present-day Miles City, Montana.[26]

In the meantime, a party of civilian road builders, led by former Iowa militia officer James Sawyers and accompanied by a military escort, was already penetrating the Powder River region. Sawyers and his men had been assigned by the War Department to establish a wagon road from Sioux City, Iowa, to Virginia City, in present-day Montana, in accordance with a congressional directive. The Sawyers expedition met a Sioux-Cheyenne war party returning home from the fighting at Platte Bridge Station on 15 August. The encounter took place near Pumpkin Buttes, west of present-day Gillette, Wyoming. After a morning battle in which the Cheyennes and Sioux were unable to overrun the Sawyers party, the Indians called for a truce. Sawyers met with the war party's leaders, including Red Cloud and the Cheyenne chief Dull Knife. George Bent served as interpreter. Bent told Sawyers that the Cheyennes demanded the hanging of John M. Chivington, the perpetrator of the Sand Creek Massacre, as a precondition for any peace treaty. He further declared that the Indians were ready and willing to fight the army and that they knew the government would withdraw its troops in the fall, leaving the Indians in control of the region. After some discussion, Sawyers agreed to give the war party food and tobacco in return for safe passage. The talks ended when two of the troops ventured out too far among the warriors. Suddenly, the Indians fired, killing them both. Sawyers's cavalry escort opened fire, and the Indians withdrew. With the aid of General Connor's forces, the Sawyers party eventually reached its destination in Montana in October after tense encounters and some fighting with Arapaho Indians enroute.[27]

On 29 August 1865, Connor's main column struck a large Arapaho village on the Tongue River, delivering a powerful blow. The general claimed that his troops killed sixty-three Indians, burned two-hundred-fifty lodges, captured five-hundred ponies, and destroyed tons of the Indians' supplies. This attack and the construction of Fort Connor (later called Fort Reno) near present-day Kaycee, Wyoming, were to be the Powder River Expedition's principal accomplishments. As George Bent had predicted, the government ordered most of its volunteer soldiers to withdraw in September, leaving a garrison at Fort Connor. The general himself was transferred to a new command in Utah.[28]

At this point, the Sioux and their allies in the north were in control of the Powder River Country. With the return of the Southern Cheyennes and Arapahos to their homelands in the fall, Red Cloud was left as the most successful war chief in the region. The symbol of his power was a magnificent tipi, the top painted blue for the sky, the bottom bordered in green for the earth, and four rainbows decorating the sides. A buffalo painted on the door meant that no one would ever go away hungry.[29]

By the end of 1865, Red Cloud was fully committed to stopping white migration and settlement in the Powder River Country and to preserving the superb hunting grounds east of the Bighorn Mountains for his own people. By doing so effectively, he had inspired like minds among the Lakotas, and from then on he was a force to be reckoned with. Military officers and other government officials, not understanding that the various Lakota bands each had their own strong leaders and that Sioux decision-making processes based on consensus often limited a chief's ability to act, soon identified Red Cloud as the prime mover behind Indian resistance to their plans for the Powder River Country. For his own part, Red Cloud did his best to fulfill his role as a protector of his people. Eighteen sixty-six would be another year of war on the Northern Plains.

5 Negotiations Fail 1866

Stand and Looks Back stated that Red Cloud fought so hard to close the Bozeman Trail because, "It was our last hunting ground. The new road cut it in two parts. The Powder River country lay in the path of this trail. This was the last good buffalo country of my nation. So you see we fought for this last pitiful remnant of our once best country."—Remington Schuyler[1]

Although warfare would resume on the Northern Plains in 1866, neither the United States government nor Red Cloud considered armed conflict to be inevitable as the year began. The resistance of the Northern Plains Indians and the long monetary drain of the Civil War fueled sentiment in Washington to seek a safe road to the Montana gold fields by treaty rather than by force of arms. Accordingly, officials laid the groundwork for a spring 1866 meeting. With the assistance of friendly Lakotas such as Big Ribs and Big Mouth, who agreed to carry messages to the winter camps of Lakota and Cheyenne bands that had resisted the army in 1865, Colonel Henry E. Maynadier, Fort Laramie's commander, invited the principal chiefs to the spring council. The colonel acted on the instructions of his superiors. Big Ribs departed the post in October 1865. Red Cloud's camp was on his list of intended stops. At the same time, however, the military ordered Colonel Henry B. Carrington and the Eighteenth Infantry Regiment west from Fort Kearny, Nebraska, to fortify the Bozeman Trail, replacing the volunteer troops that had previously garrisoned it. While the government may have expected the council to result in Sioux and Cheyenne acquiescence to the trail's continued use, it did not work out that way.[2]

While Carrington's troops prepared to move out, Fort Laramie buzzed with activity. At noon on 8 March, Spotted

Tail and his band of Brules reached the post bearing the body of his recently deceased daughter, who had expressed the wish of being buried there. Colonel Maynadier made arrangements for an elaborate funeral, in which mourners placed the remains on a scaffold in accordance with Lakota tradition northwest of the fort. Maynadier had made a friend, and Spotted Tail would soon agree to peace terms.[3] Three days later, the Brule chief escorted Red Cloud, who was accompanied by one hundred fifty Oglala warriors to Fort Laramie. Because army and Indian Office officials saw Red Cloud as the principal leader of the resistant Sioux, the government wanted his participation in the treaty process.

Red Cloud's willingness to negotiate in the spring of 1866 was likely due to difficulties his people had just endured over the recent winter. Hunting in the Powder River Country had been poor, and game had been scarce. On 24 March, Colonel Maynadier wrote to Commissioner of Indian Affairs Dennis N. Cooley that Red Cloud's people had survived "a near approach to starvation."[4]

Colonel Maynadier and Vital Jarrot, Indian agent for the Upper Platte, met with Red Cloud and Spotted Tail. To impress the chiefs, Maynadier and Jarrot used the telegraph to allow Edward B. Taylor, head of the peace commission that would come to Fort Laramie, to communicate with them.[5] It may have been the first time that a representative of the United States government sent a diplomatic message to Indian leaders by telegraph. Although Red Cloud was initially reluctant to enter the telegraph office, he finally agreed. Taylor sent this message:

"The Great Father at Washington [the president] has appointed Commissioners to treat with the Sioux, the Arapahos, and Cheyennes of the Upper Platte on the subject of peace. He wants you all to be his friends and the friends of the White man. If you conclude a treaty of peace, he wishes to make presents to you and your people as a token of his friendship. A train loaded with supplies and presents cannot reach Fort Laramie from the Missouri River until near the first of June and he desires that about that time be agreed upon as the day when his commissioners all meet you to make a treaty."[6]

Red Cloud and his followers seemed to accept the message, and the chief said that he would stay in the area until the commissioners arrived. Taylor's telegram did not mention the Powder River route, nor apparently did Maynadier or Jarrot in their own discussions with Red Cloud at any time before the peace council convened on 5 June.[7] In the light of the events that followed, the government representatives' decision not to inform Red Cloud of their intent to keep the Bozeman Trail open backfired.

Taylor and his fellow commissioners arrived at Fort Laramie on 30 May 1866 and began organizational work on 1 June. Over the next several days, commissioners met many of the chiefs and headmen of the assembled tribes.[8] Red Cloud made a formal entrance to the post with a company of about fifty to sixty handpicked warriors, clearly intending to steal the scene. One soldier described Red Cloud and his men:

Their peculiar equipment and trappings, loose and dangling, the parti-colored blankets, the quivers hiding their bows and arrows, hung free on their backs; their long spears glistening and catching the sun's reflection; riding without any particular formation — somewhat scattered, careless, if not reckless — they made an attractive spectacle. All made up a harmony of color, a jingling of metal accouterments, a dash, a swing, a movement that held one to a feeling that thrilled. And keeping up that swinging gait, sitting their horses like centaurs, they dashed in on the parade ground with a confidence and fearlessness that was truly admirable.[9]

At 10 o'clock on 5 June, the council began in front of the post headquarters with nineteen chiefs present. Taylor's opening remarks explained the government's goals for the proposed treaty. Taylor assured his audience that no purchase of the Indians' homeland was contemplated. Rather, the government sought their permission to build roads through the Powder River Country for public business and emigrant access to the mining districts of the West. Taylor frankly admitted that the white presence would scare off some game and stated that the government would compensate the tribes for the damage done.[10]

Taylor ended his speech with a summary of the government's position: "The Great Father does not wish to keep many soldiers in this country. He will be satisfied if the mails and telegraph are not molested. We do not ask you to give up this country, or sell it. We only ask for roads to travel back and forwards, and no roads will be made except by orders from the Great Father, so as not to disturb the game and whatever damage are done by roads would be paid for by the Great Father."[11] According to a Denver newspaper, Red Cloud shook hands with the commissioners after Taylor's speech and remarked that "he had nothing to say as he [Red Cloud] wished to get all the Indians together, and have a talk."[12]

Red Cloud, Spotted Tail, Red Leaf, and Man Afraid of His Horses spoke at the next day's council session. Red Cloud was to the point: "Yesterday we talked about small matters, and today we want to talk about big matters. My people on both sides of the road have [only] bows and arrows, and we came in here and told you all we wanted. We are 21 bands. All we want is peace. We have come here for you to give us instructions that we may live."[13] Taylor's report of the council proceedings in a letter to the commissioner of Indian affairs on 9 June appears disingenuous when compared with other eyewitness accounts. Taylor claimed that the chief's speeches had been "marked by moderation and good feeling," which was certainly not true of Red Cloud.[14] In contrast, the *Salt Lake City Union Vedette*'s correspondent at Fort Laramie reported that during the session the Indians had strongly objected to the whites transiting their country or making new roads and asked that the troops be withdrawn from Fort Reno. The peace commissioners, however, were equally determined to conclude a treaty of peace that "allowed [whites] to go wherever they see fit."[15] According to Taylor, the Indian leaders at the conclusion of the session had asked for time to bring in those of their people who were encamped on the headwaters of the White River, some fifty or sixty miles north of Fort Laramie. Red Cloud and his entourage headed north as Taylor wrote his report.[16]

On 13 June, Colonel Carrington and his command reached Fort Laramie.[17] Red Cloud soon learned of the presence of Carrington's eight companies of infantry on their way to fortify the Bozeman Trail. For Red Cloud, the implication was clear. The whites had lied; the Great Father did wish to bring soldiers to this country.

There are conflicting accounts of Red Cloud's specific words and actions upon learning of Carrington's mission. Most historians report that Red Cloud returned to Fort Laramie to lecture the commissioners and then make a dramatic exit on 13 June. However, two newspaper correspondents who were present at Fort Laramie stated that Red Cloud was still absent from the post on 16 June.[18] What is beyond dispute is that Red Cloud's reaction to Carrington's presence was negative, and he withdrew from the peace conference in disgust.

Later, when Black Horse urged Red Cloud to reconsider and take what the white man offered rather than fight, Red Cloud is said to have replied: "White man lies and steals. My lodges were many, but now they are few. The white man wants all. The white man must fight, and the Indian will die where his fathers died."[19] Red Cloud's War was about to begin.

6 Red Cloud's War 1866–1868

The Indian was assailed in his last covert, on the only soil where game remained, and it was understood by him to be, as it was in fact, his final struggle for independence and self-support, after the manner of his fathers. —Colonel Henry B. Carrington[1]

For over two years Red Cloud, his warriors, and their allies waged war against all comers on the Bozeman Trail, with the intent of closing the Powder River Country to Euro-American invaders. While the United States Army built and maintained Fort Reno, Fort Phil Kearny, and Fort C. F. Smith to protect white travelers on the route from Fort Laramie to Montana, the troops garrisoning the three posts became virtual prisoners. Brigadier General Patrick E. Connor's Powder River Expedition had promised to solidify the Bozeman route in 1865, but Indians attacked many wagon trains traversing the region in 1866 causing significant loss of life, both military and civilian. Describing the situation in its edition of 1 March 1867, the *Salt Lake City Daily Union Vedette* credited Red Cloud with almost superhuman powers as leader of the Indian resistance: "Whenever a white made his appearance on the Powder River road, anywhere of its five hundred miles in length, he was sure to meet the seemingly omnipresent Red Cloud, on his gray horse. [Wagon] trains were surprised by night and ambushed by day. Government herds were stampeded at every point. Troops were surprised and slaughtered by scores. Communication between posts was cut off. Wood and hay parties driven in, and at last the military posts were in a state of siege."[2] The *Union Vedette* report, while exaggerated in some respects, accurately describes Red Cloud's tactics. Stealth, swift movement, and surprise attacks designed to hurt and harass the enemy while exposing the war party to minimum risk were hallmarks of the Plains Indian military tradition. Ambush tactics also made sense because most Lakota and allied Indian warriors lacked up-to-date firearms, and many still depended on bows and arrows, lances, knives, tomahawks, or war clubs. Plains Indians generally avoided risky assaults on well-fortified enemies.[3]

By one account, Red Cloud's reluctance to attempt a direct attack on Fort Phil Kearny stemmed in part from information gathered by Cheyenne Indians who had been inside the walls on 16 July for a conference with Colonel Henry B. Carrington, the post commander. According to the Cheyenne warrior Two Moon, the stockade was well protected, with blockhouses commanding the corners and artillery further bolstering the defenses. Two Moon reportedly suggested to Red Cloud that the Indians should attack isolated parties of soldiers away from the forts, such as wood-gathering details. Red Cloud agreed. Lakota warriors, together with their Cheyenne and Arapaho allies, continually harassed work parties outside the three Bozeman Trail forts. Red Cloud further reasoned that if the Indians could cut off the forts' communications and supplies during the winter, the soldiers could be starved out.[4] While Red Cloud and his allies were too crafty to assault the forts head-on, their occupants did not dare venture beyond the walls without a strong guard. One report claimed that the Fort Phil Kearny garrison endured fifty-one hostile contacts with Indians around the post from 26 July to 21 December 1866. During this period fifty-eight civilians and ninety-six soldiers died, most of the latter with Captain William J. Fetterman on 21 December.[5]

The Fetterman fight was both the largest single engagement and the Indian warriors' most successful use of decoy-and-ambush tactics during Red Cloud's War. In this battle Lakota, Arapaho, and Cheyenne warriors annihilated Fetterman's command of eighty men. On the morning of 21 December, the Indians attacked a wood-cutting detail.

In response, Colonel Carrington sent out a relief column under Captain Fetterman. A ten-man decoy force led by Crazy Horse lured Fetterman away from the woodcutters. Despite Carrington's strict orders to the contrary, the troops followed over a ridge that blocked the view of the fort's lookouts and prevented the use of its artillery. Approximately fifteen hundred Indians lay in ambush over the ridge, overwhelming Fetterman and his men in minutes. Years later, Red Cloud told James H. Cook that some of the soldiers had apparently been paralyzed by fear and offered little resistance. The officers, however, tried to rally their men and fought bravely to the end. The chief estimated that only eleven Indians had died on the battlefield, but a number of others later succumbed to their wounds.[6]

Although Euro-Americans considered Red Cloud to be the prime mover behind what they called the Fetterman massacre, chiefs of the Minneconjou Lakotas had taken the initiative in this particular case. War leader High Back Bone (also known as Hump) served as principal organizer of the battle. Oglala, Brule, Two Kettle, Hunkpapa, Cheyenne, and Arapaho warriors joined the Minneconjous by invitation. While some Minneconjous in later years claimed that the Oglala chief had missed the Fetterman fight, Red Cloud maintained to the end of his life that he had been a participant. Regardless of his specific whereabouts on the day Fetterman's troops fought and died, Red Cloud remained the most influential Lakota leader opposed to the white presence on the Bozeman Trail.[7]

After the Fetterman fight, the Indians dispersed into smaller camps, as was their usual practice in winter. The Lakotas and their allies resumed hit-and-run raiding along the Bozeman Trail in the spring of 1867. None of the many hostile encounters between Indians and soldiers that year approached the Fetterman fight in size until August, when the last two large engagements of the war took place. On 2 August, Red Cloud's war party attacked a detachment of soldiers guarding a wood-cutting detail about five miles north of Fort Phil Kearny. The encounter is known as the Wagon Box Fight, after the soldiers' makeshift fortification constructed of wagon boxes with the wheels and chassis removed. During this encounter, the Lakota war party was surprised to find that the troops had a powerful new weapon — the Model 1866 Springfield breech-loading rifle, which had a much greater rate of fire than the muzzle-loading weapons previously carried by most infantry soldiers. Consequently, Captain James Powell's group of two officers, twenty-nine enlisted men, and six civilians fought Red Cloud's warriors to a standstill despite being heavily outnumbered. The Indians withdrew at the sight of a relief column from the fort.[8] Red Cloud later referred to the Wagon Box Fight as the battle of "the much talk gun."[9]

Red Cloud's Cheyenne allies to the north also learned of the army's new weapon from firsthand experience. The day before the Wagon Box Fight, a Cheyenne war party accompanied by a few Lakotas had attacked a hay-cutting detachment a few miles from Fort C. F. Smith in present-day Montana. In this struggle, known as the Hayfield Fight, one officer, twenty-three enlisted soldiers, and twelve civilians stood off their attackers for half a day. As in the Wagon Box Fight, the Indians had the advantage of much greater numbers, but the defenders used their new breech-loading rifles to good effect and all but three survived.[10]

After the Hayfield and Wagon Box fights, Lakota and allied warriors avoided large-scale attacks on sizable, well-armed military detachments. The Indians, however, still seized opportunities to raid wagon trains, run off livestock, and kill soldiers and civilians who strayed too far from safety, either as individuals or in groups too small to defend themselves. The Lakota objective of closing the Bozeman Trail did not require the annihilation of large army units. Convincing the whites that the route was too difficult to keep open would suffice. Indeed, records indicate that only one emigrant train traveled the main route of the Bozeman Trail in 1867. The modern label "guerrilla warfare" fittingly describes Red Cloud's War, as it does the Plains Indian wars in general.[11]

Red Cloud did not rely solely on force of arms to achieve his goal of removing the white presence in the Powder River Country; he also employed diplomacy, even to the point of negotiating with traditional enemies of his people. Soon after Red Cloud abandoned the Fort Laramie peace talks with the whites in June 1866, he and Man Afraid of His Horses contacted the Crows, the very people the Oglalas had fought for possession of the region and the adversaries against whom Red Cloud had earned much of his warrior reputation. Red Cloud and Man Afraid offered the Crows peace with the Oglalas in exchange for an alliance against the whites. While the Crow chiefs apparently could not bring themselves to ally with the Lakotas, officers at Fort C. F. Smith learned from visitors in August 1866 that some younger Crow men favored the proposed alliance because they resented the Bozeman Trail and the threat the white presence posed to the Montana hunting grounds still in Crow hands. Crow chiefs who visited Fort Laramie to negotiate with the government in November 1867 made it clear that their people were opposed to the Bozeman Trail as much as the Lakotas, who had apparently granted the Crow party safe passage through the lands under their control.[12]

By the winter of 1867–1868, Red Cloud's willingness to negotiate with his people's enemies also extended to the United States government. The commissioners who met the Crows at Fort Laramie had a congressional mandate to negotiate peace with all the Plains tribes. While Red Cloud did not meet the commissioners in person, he sent them a message through intermediaries. Red Cloud's negotiating position was clear. In the words of the commission's official report of January 1868: "[Red Cloud's] war against the whites was to save the valley of the Powder river, the only hunting ground left to his nation, from our intrusion. He assured us that whenever the military garrisons at Fort Phil. Kearney [sic] and Fort C. F. Smith were withdrawn, the war on his part would cease."[13]

Because of the difficulty in providing safe passage for emigrants over the Bozeman Trail, the impending completion of the Union Pacific Railroad, and the expected development of rail routes to the Montana mining region that would bypass the Powder River Country, officials in Washington decided to yield to Red Cloud's key demand in early 1868. In accordance with the decision, General Ulysses S. Grant, the army's senior officer, directed Lieutenant General William T. Sherman to abandon the Bozeman Trail forts on 2 March. The actual withdrawal took several more months, however. The army's Department of the Platte, field headquarters for the troops in the Bozeman Trail forts, did not issue its own withdrawal order until 19 May. The Fort C. F. Smith garrison left on 29 July, and Forts Phil Kearny and Reno were finally abandoned in August.[14]

While the military began its slow withdrawal from the disputed forts, the peace commissioners prepared to meet with prominent chiefs at Fort Laramie in the spring of 1868. On 30 March, couriers from Fort Laramie headed for the northern Indian camps to spread the word of the government's decision to abandon the three forts on the Bozeman Trail and hold a peace council. Lieutenant Colonel Adam J. Slemmer, commanding Fort Laramie, had reported to his superiors on 5 March that Red Cloud was encamped with about eight hundred lodges on the Powder River waiting for the Bozeman Trail forts to be evacuated.[15]

The peace commissioners reached Fort Laramie about 9 April carrying a prepared treaty, with no expectation that Indian input would change its terms. No Indians were present when the commissioners arrived, but Spotted Tail's band of Brules and the so-called Laramie Loafers (Lakotas who lived in the vicinity of the fort and had not been at war) appeared and signed the treaty on 29 April. These signatures were easy to obtain, for Spotted Tail had accepted the government's 1866 proposals, which Red Cloud had rejected.[16]

The treaty presented for Spotted Tail's signature at the 29 April council was a major milestone in the government's relations with the Sioux tribes. Forts Reno, Phil Kearny, and C. F. Smith along the Bozeman Trail were to be abandoned. The accord established a Great Sioux Reservation encompassing the western half of what is now South Dakota, including the Black Hills, and a narrow strip of land in present-day North Dakota. According to the treaty, the reservation was to be "for the absolute and undisturbed use of and occupancy of the Sioux." However, the document contained a crucial proviso: "They [the Indians] will not in the future object to the construction of railroads, wagon roads, mail stations, or other works of utility or necessity, which may be ordered or permitted by the laws of the United States."[17] This provision helped set the stage for the eventual dispossession of the Sioux tribes.

Another stipulation that had a surface appeal to potential Indian signers was the clause that set aside the region between the Black Hills and the Bighorn Mountains — including the Powder River Country — as "unceded" territory, where reservation Indians were free to hunt so long as buffalo remained sufficiently numerous. The Sioux tribes retained similar hunting rights north of the North Platte River and on the Republican River. After game disappeared, the Indians would not have the right to reside permanently in these areas. The land would revert to the public domain and could then be made available to settlers. Here, the treaty-makers again opened the door for white expansion.[18]

In order to encourage the Sioux tribes to give up their traditional way of life and take up agriculture on the Euro-American model, the government agreed to supply those Sioux who would permanently move to the reservation with food rations and annual clothing allotments for the next thirty years. A variety of specialists, services, infrastructure, and equipment was also promised to reservation Indians. An agency, or reservation headquarters, was to be created on the Missouri River and staffed by an agent, teachers, a doctor, a blacksmith, a miller, and an experienced farmer. Rations were to be sent to the Missouri River agency, not to locations such as Fort Laramie where the Sioux tribes had been accustomed to trading with whites. Finally, the treaty provided that three-fourths of the reservation's adult male population had to approve future cessions of any reservation lands.[19]

Spotted Tail's Brule band and the Laramie Loafers were the first Indians to sign peace agreements with the United States at Fort Laramie in 1868. Crow representatives who arrived on 1 May were shocked that the treaty signed by Spotted Tail two days earlier had already reserved hunting rights in the Powder River Country for the Sioux. On 7 May, the Crows signed a separate treaty that placed their reservation in southern Montana. Other Lakota bands, as well as Yanktonai Sioux, arrived at the fort and affixed their signatures to the Treaty of Fort Laramie, including Red Cloud's fellow Oglala leader Man Afraid of His Horses and a number of Minneconjou chiefs on 25 May. After the last of the peace commissioners left Fort Laramie in late May, the duty of collecting and witnessing further Indian signatures fell to the post commander and his officers.[20]

The commissioners departed without the one signature they had deemed most important—that of Red Cloud. This fact concerned the *Cheyenne Leader*, whose issue for 6 June 1868 editorialized, "The fact is, Red Cloud is the supreme ruler of all the Indians between the Platte and the Mississippi rivers, and though all but he might sign treaties, one word from him for war would dispel all their recollections of promises of peace." For his own part, the *Leader* reported, Red Cloud's messages to officials at Fort Laramie made it clear that he would not sign any treaty until the Bozeman Trail forts were actually abandoned.[21] The day before the *Cheyenne Leader* ran its article on Red Cloud, however, Lieutenant Colonel Slemmer notified his superiors that two parties of Lakotas had just visited Fort Laramie and signed the treaty. All but one of the headmen who signed at that time belonged to Red Cloud's own band of Oglalas. They told the post commander that Red Cloud was organizing a war party against the Shoshonis to avenge the death of a son, and he would sign the treaty afterwards. Slemmer optimistically reported, "Everything looks very favorable for a lasting peace with the Sioux nation."[22]

Despite Slemmer's optimism in June, Red Cloud took his time. He was in the vicinity of Fort C. F. Smith when that post was abandoned on 29 July, and warriors from his band marked the occasion by burning the vacant fort. The last of the Bozeman Trail posts, Fort Reno, was evacuated on 18 August, and still Red Cloud did not come to Fort Laramie. Apparently, the next immediate priority for Red Cloud and his people was to obtain meat for the winter. A group of Cheyennes visiting Fort Laramie in September told Lieutenant Colonel Slemmer that Red Cloud and his band were hunting buffalo near the mouth of the Rosebud River.[23]

After the fall buffalo hunt, Red Cloud left his camp on the southeast branch of the Belle Fourche River and headed for Fort Laramie. Leading an impressive group of prominent chiefs, headmen, and warriors, he arrived at about ten o'clock on the morning of 4 November. The delegation included about one hundred twenty-five representatives of the Hunkpapa, Blackfeet, Brule and Sans Arc Lakotas in addition to his own Oglalas. Some chiefs who had already signed the treaty, such as Red Leaf, Brave Bear, and Man Afraid of His Horses, accompanied Red Cloud. Major William M. Dye, the new post commander due to the death of Lieutenant Colonel Slemmer on 7 October, greeted the Lakota delegation.[24]

Major Dye began the three-day peace council with a large feast for the Lakota delegates. When introduced, Red Cloud remained seated and offered only his fingertips to those who shook his hand, leading Euro-American witnesses to characterize his actions as arrogant. Red Cloud's intentions were likely misconstrued, as it was customary among his people for peace negotiators to extend the fingers to show that they did not hold a weapon. Once substantive talks began, Red Cloud asked questions about Major Dye's authority to negotiate the treaty, knowing that Dye had not been among the commissioners who had initially signed it. Dye replied that as post commander he had power to sign the treaty and witness the Lakota delegates' signatures.[25]

Once satisfied as to the major's authority, Red Cloud queried Dye about the treaty's provisions point by point. When Dye began to explain the document's clauses directing the Lakotas to move to the reservation and take up agriculture for their livelihood, Red Cloud interrupted him. The chief declared that he knew about those terms, but he and his people did not want to leave their traditional lands or give up hunting to become farmers. What he did want, he said, was some powder and lead to fight the Crows. According to Red Cloud, his enemies had obtained such supplies from the peace commissioners, and it was only fair for his people to get some as well. Dye responded that he could not issue powder and lead to Indians who were at war, adding that Brevet Major General William S. Harney, who had been one of the peace commissioners, was the proper authority to deal with Red Cloud's request.[26]

Major Dye explained that when Congress had appropriated money to feed and clothe the Sioux tribes as part of the peace process, it had bypassed the civilian Office of Indian Affairs and directed the army to administer the funds. Thus, a military district commanded by General Harney would run the program for the new Sioux reservation on the Missouri. Dye promised that the government would supply all the Indians' basic needs, but they must move to the reservation to receive these benefits. At the close of the first council session, Dye asked Red Cloud to think about the matter overnight.[27] That evening, Ann Vogdes, an officer's wife, entertained Red Cloud and his fellow chiefs Red Leaf and Big Bear. Vogdes recorded in her journal that her guests "sat themselves down in rocking chairs and on sofas, with as much ease and grace, as if they had been born there, and knew no other life."[28]

In subsequent sessions, Major Dye reviewed the treaty provisions, highlighting portions that benefited the Indians. Red Cloud renewed his plea for powder and lead. Dye telegraphed his superiors about the matter at the chief's request. When he received a telegram in reply from Department of the Platte commander Christopher C. Augur in Omaha, Nebraska, the major read it to the council. Augur's telegram told Red Cloud that the Oglalas would have to receive their supplies from General Harney. Red Cloud followed with more questions, focusing on the extent of the Lakotas' territory and the treaty provision barring whites from it. Finally, on 6 November 1868, Red Cloud washed his hands with dust from the floor to signify his reluctance and put his mark on the document. After his fellow Lakota chiefs signed the treaty, Red Cloud asked all the Euro-American witnesses to touch the pen.[29]

After the signing, Red Cloud shook hands with his white counterparts and then gave a long speech, in which he pointed to the opening of the Bozeman Trail without Indian consent as his cause for war. Now that the whites had abandoned it, he saw no further reason for fighting. The chief warned that young Lakota warriors might be difficult to control, but he himself would honor the agreement as long as the whites did. Evidently considering the "unceded" territory described in the treaty as a temporary refuge for his people, Red Cloud was noncommittal about going immediately to the reservation for rations and presents. He would

have to see the goods before deciding what to do. Remark-
ing that he still wanted to visit Fort Laramie occasionally,
Red Cloud expressed the hope that its soldiers would treat
his people with courtesy and that fur traders already familiar to the Lakotas could resume business there. The chief
insisted that the government should supply the Lakotas
with lead and powder to facilitate their hunting due to the
scarcity of game. In closing, Red Cloud announced plans to
go to the Powder River, along with Man Afraid of His Horses,
Brave Bear, and their bands, to hunt buffalo and fight the
Crows.[30]

The treaty had two immediate results for Red Cloud.
One was widespread recognition (at least among Euro-
Americans) of the Oglala leader as the principal chief of
all the Sioux tribes due to his ability to close the Bozeman
Trail forts. Writing on 5 December to Brevet Brigadier General George D. Ruggles, an official at the Department of the
Platte's headquarters, Major Dye stated, "Red Cloud is now
looked on by the Sioux Nation as their Big Chief and that
until he signed the treaty they did not look upon the war
as closed." Dye cited the Oglala leader Brave Bear as the
source of his information. Furthermore, Dye declared, "If
Red Cloud occupies the high position Brave Bear says he
does we will hereafter look to him principally for the obser-
vance by Indians of their treaty obligations."[31] Like Major
Dye, federal officials in Washington regarded Red Cloud as
paramount chief of the Sioux.[32]

Even so, signing the treaty cost Red Cloud some influence among his own people. As he himself had warned in
his post-signing speech, Red Cloud could not guarantee
that all young Lakota warriors would accept the Treaty of
Fort Laramie. Previously, he had been the ideal war chief,
demonstrating resolute leadership in the campaign to close
the Bozeman Trail. Now, an element of his people would
denounce his accommodation with the government, regarding those holdouts who disdained any relations with

the whites as heroes. Thus, Lakotas who opposed the treaty ceased to recognize Red Cloud's authority and looked to other leaders, such as Crazy Horse and Sitting Bull. Indeed, Crazy Horse would emerge as the symbol of Lakota military resistance to the United States government over the next decade, while Red Cloud would take the path of defending his people's interests by other means.[33]

During the summer of 1868, a large group of Oglalas camped some forty miles northwest of present-day Lusk, Wyoming. One day, horsemen circled the camp, summoning the people to a council at which tribal elders appointed four distinguished warriors as *Ongloge Un*, or shirt-wearers: Young Man Afraid of His Horses, Crazy Horse, Sword, and American Horse. While the appointment was a great honor, it also carried important responsibilities. Shirt-wearers kept the peace in camp, resolved disputes, and ensured that the rights of all people in the community were respected. Appointed for life, shirt-wearers were expected to maintain high ethical standards. Protectors of the general welfare, they wielded great power in the Oglala political structure.[34]

Some historians have puzzled over the fact that Red Cloud was not appointed a shirt-wearer, while others point to this alleged snub as evidence that he had lost significant power. In fact, the selection of the shirt-wearers did not diminish Red Cloud's standing among his people. He was now the most influential tribal chief among the Lakotas. The shirt-wearers were all young men of robust health who had proven themselves in combat; they were expected to lead in the future. Red Cloud's role was to continue to lead in the present.

Some observers, such as John R. Brennan, Indian agent at the Pine Ridge reservation in the early twentieth century, pointed to Red Cloud's War and the treaty that ended it as unqualified triumphs for the Sioux because of the closure of the Bozeman Trail and the removal of the forts that garrisoned it. Writing just after Red Cloud's death in 1909,

Brennan claimed the conflict of 1866–1868 had been "the only instance in the history of the United States in which the government has gone to war and after war has made peace, conceding everything demanded by the enemy and exacting nothing in return."[35] More recent writers see the outcome quite differently. The 1868 Fort Laramie treaty was an undeniable strategic victory for the whites because it set the stage for the eventual dispossession of the Sioux. As one historian wrote in 1976, the treaty created a "prison without walls" for the Lakotas.[36] Red Cloud would spend the remainder of his days as chief attempting to ameliorate Euro-Americans' impacts on his people.

7 Finding a Place 1869–1873

His name has been heralded with electric speed, within a month, to the remotest parts of the civilized world, and his position has made him in name, if not in reality, great. —Omaha Weekly Herald, *1 June 1870*[1]

Following the signing of the Treaty of Fort Laramie, Red Cloud returned north. In February 1869, he was reportedly camped near the remains of Fort Phil Kearny, which warriors had burned to the ground after the garrison's departure. Major William M. Dye, commanding Fort Laramie, received a message from the chief stating that he intended to trade at that post in the spring, as had been common practice for his people before the recently concluded war.[2] Red Cloud's message gave no indication that he was in any hurry to receive his people's promised annuities and food rations on the Missouri River or take up permanent residence on the Great Sioux Reservation as envisioned in the treaty. Indeed, by both his words at the November 1868 treaty signing and his actions afterward, Red Cloud made it clear that although he had given up resisting the United States government by force of arms, he nonetheless intended to maintain traditional Lakota life as much and as long as possible. Other Lakota leaders who had accepted the treaty shared these intentions, which ran counter to the government's stated goal to transform the Sioux tribes from roaming hunters into settled agriculturalists on the Euro-American model. For that process to begin, federal officials needed to move the tribes onto their designated reservations.[3] As events were to demonstrate over the five years following the treaty signing, moving Red Cloud's Oglalas to the Great Sioux Reservation was easier said than done.

50

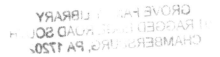

As promised, Red Cloud came to Fort Laramie in the spring of 1869. His arrival, in which the chief showed that he had not lost his warrior skills, was both startling and embarrassing to the post commander. One morning in late March, the commander awoke to find that Red Cloud, Man Afraid of His Horses, and Red Leaf had led hundreds of mounted warriors to form up on Fort Laramie's parade ground while their families watched from a distance. "Fort Laramie was not only surprised but astounded," the *Wyoming Weekly Leader* later recorded. Major Dye rushed to arrange his own show of force, calling out the entire garrison and preparing his artillery for use. After a standoff of several hours in which neither side moved from its position, Dye sent an interpreter to learn the purpose behind Red Cloud's "mysterious maneuvers." He returned with the terse message from the chief, "We want to eat."[4]

During the negotiations that followed, Red Cloud stated that his people needed food badly. As the *Weekly Leader* observed, the warriors could easily have attacked the post with devastating results, but their chief had not come to fight. Instead, Red Cloud's people made camp just outside the post. When asked why he did not go to the Missouri River to trade as the 1868 treaty stipulated, Red Cloud replied that he "didn't belong on the Missouri," adding that he expected to trade at Fort Laramie as he always had done in the past. To reduce tensions, Red Cloud agreed to keep his warriors in camp and leave the actual trading to the Lakota women. According to the newspaper's account, the women had many ponies to trade but few buffalo robes and skins.[5] Despite Major Dye's attempt to persuade them to go to the Missouri, Red Cloud and his followers spent the summer of 1869 in Wyoming's Wind River Valley.[6]

In April 1870, the commanding officer at Fort Fetterman, a few miles northwest of present-day Douglas, Wyoming, reported that Red Cloud wanted to visit Washington, D.C., to discuss the Treaty of Fort Laramie in person with President Ulysses S. Grant. Some Euro-American observers suggested that John Richard, Jr., a mixed-blood trader whose father, John Richard, Sr., had married Red Cloud's niece, had influenced the chief to make this request. The family relationship made the Richards Red Cloud's people. The younger Richard was in legal trouble, facing a murder charge for fatally shooting a Fort Fetterman soldier while in a drunken rage on 9 September 1869, perhaps in a dispute over a woman. He fled the post and took refuge among his Oglala relatives. In December, newspapers in Cheyenne, Wyoming, and Omaha, Nebraska, ran stories claiming that John Richard, Jr., was plotting to incite the Crow and Sioux tribes to attack white settlers. Richard's white and Indian friends at the Whetstone Agency near Fort Randall in present-day South Dakota petitioned President Grant for a pardon on his behalf in January 1870, suggesting that his services could be useful in maintaining peace with the Plains tribes. How much the younger Richard actually influenced Red Cloud's desire to visit Washington remains unknown. However, federal authorities allowed Richard to accompany Red Cloud on the trip as an interpreter and blocked his prosecution for the soldier's death after the Lakota delegation returned home.[7]

President Grant approved Red Cloud's request to visit Washington on 3 May 1870. The chief and his fellow Oglala delegates arrived at Fort Fetterman on 16 May to begin their historic journey. Several hundred Oglalas were on hand to see them off. In addition to Red Cloud, the group included Brave Bear, his son Sword, Red Dog, Yellow Bear, High Wolf, Sitting Bear, Little Bear, Long Wolf, Bear Skin, Brave, Afraid, Red Fly, Rocky Bear, Swing Bear, Black Hawk, and The One That Runs Him Through. Two days later, the delegates arrived at Fort Laramie, where Colonel John E. Smith, who had been the final commander of Fort Phil Kearny, was waiting to escort them. The Indians' hand-picked interpreters, including a smug John Richard, Jr., accompanied the party, which left Fort Laramie on 26 May. The delegation boarded its train at Pine Bluffs, Wyoming, in order to avoid potentially hostile citizens in Cheyenne, the territorial capital. The group arrived safely in Washington on the first of June.[8]

Commissioner of Indian Affairs Ely S. Parker and Secretary of the Interior Jacob D. Cox met with Red Cloud and his fellow Lakota delegates for the first time on 3 June. Red Cloud made his hosts uncomfortable by immediately asking them to send food, ammunition for hunting, and a telegram announcing his safe arrival in Washington to his people at home. Cox agreed to send the telegram but deferred action on the chief's other requests and left the meeting, saying he needed to see President Grant. Commissioner Parker then told the delegates that he had arranged a tour of the city for them. After visiting such places as the Capitol, an arsenal, and the Washington Navy Yard on 3 and 4 June, the Lakota delegation attended a lavish state dinner hosted by President Grant at the White House on the sixth. As was standard practice when dealing with Indian delegations, government officials had chosen these activities to impress their guests with the Great White Father's power and splendor.[9]

When substantive talks resumed in a meeting with Secretary Cox on 7 June, Red Cloud showed no sign of having been awed into compliance with the government's wishes. He asked Cox to "tell the Great Father to move Fort Fetterman away" and declared that he wanted "Great Father to make no roads through" the Black Hills and the Bighorn Mountains. Further, Red Cloud stated, "I don't want my reservation on the Missouri. This is the fourth time I have said no." The chief also complained that "only a handful" of the goods the government sent to his people actually arrived, the rest having been stolen enroute by corrupt middlemen.[10] On 9 June, the Lakota delegation finally got its chance to speak directly with President Grant. Red Cloud pressed Grant to agree to abandon Fort Fetterman, but

Grant declined, saying the fort was needed to protect Indians and whites alike.[11]

Despite this disappointing response from President Grant, Red Cloud and his fellow delegates met again with Secretary Cox the next day, 10 June. Red Cloud created a furor with an angry response to Cox's discussion of the 1868 treaty, declaring, "I never heard of it and do not mean to follow it." Red Cloud argued that the document he had approved at Fort Laramie only provided for an end to fighting with the whites in exchange for the closure of the three Bozeman Trail forts. Cox asked Red Cloud to take a copy of the treaty so it could be explained to him, but the chief refused, saying, "It is all lies."[12] Red Cloud and his fellow Lakota delegates did not accuse the peace commissioners of having lied in 1868; rather, the Indians suggested that the interpreters at the Fort Laramie peace council had misled them. Not wanting the talks to collapse, officials persuaded the Lakota delegates to return to the Interior Department office for another session on 11 June, a Saturday.[13]

The next day, Secretary Cox announced that the government would make certain concessions. The Oglalas could live temporarily on the headwaters of the Cheyenne River, northeast of Fort Fetterman in the "unceded" territory where the Lakotas retained hunting rights under the 1868 treaty. While the government still expected Red Cloud's people to trade at the Missouri River, the Indians would not have to receive their annuity goods there. The government asked the Oglalas to submit the names of the men they wanted for their agent and traders to Commissioner Parker. On 12 June, Red Cloud responded that he disapproved of military men as agents because they frightened his people and that poor men appointed as agents would be tempted to steal the Indians' annuities. The chief suggested Benjamin B. Mills for agent and William G. Bullock for trader. Red Cloud already knew and trusted both men. Cox replied that he was not yet ready to name agents and traders, but that honest men would be appointed. While Red Cloud's closing speech showed that he was not satisfied with Cox's answer, he still shook hands with Cox and Parker when the meeting ended.[14]

With their business in Washington completed, Red Cloud and the other Oglala delegates traveled on 14 June to New York City, where he and Red Dog were scheduled to speak on the sixteenth at Cooper Union, a college founded by industrialist Peter Cooper. Although Red Cloud had not wanted to come to New York, the auditorium was packed for his speech. The chief spoke one sentence at a time to an interpreter, who translated Red Cloud's words for the Rev. Howard Crosby, who then relayed them to the audience.[15] Red Cloud's message was consistent with the position he had taken in his meetings with President Grant, Secretary Cox, and Commissioner Parker:

We came to Washington to see our Great Father that peace might be continued. The Great Father that made us both wishes peace to be kept; we want to keep peace. Will you help us? In 1868 men came out and brought papers. We could not read them, and they did not tell us truly what was in them. We thought the treaty was to re-move the forts, and that we should cease from fighting. But they wanted to send us traders on the Missouri. We did not want to go to the Missouri, but wanted traders where we were. When I reached Washington the Great Father explained to me what the treaty was, and showed me that the interpreters had deceived me. All I want is right and just. I have tried to get from the Great Father what is right and just. I have not altogether succeeded. I represent the whole Sioux nation, and they will be bound by what I say. . . . Look at me. I am poor and naked, but I am the Chief of the nation. We do not want riches but we want to train our children right. Riches would do us no good. . . . The riches that we have in this world, Secretary Cox said truly, we cannot take with us to the next world. Then I wish to know why Commissioners are sent out to us who do nothing but rob us and get the riches of this world away from us? I was brought up among the traders, and those who came out there in the early times treated me well and I had a good time with them. . . . But, by and by, the Great Father sent out a different kind of men; men who cheated and drank whisky; men who were so bad that the Great Father could not keep them at home and so sent them out here. I have sent a great many words to the Great Father but they never reached him. They were drowned on the way, and I was afraid the words I spoke lately to the Great Father would not reach you, so I came speak to you myself; and now I am going away to my home.[16]

The Cooper Union audience greeted nearly every sentence of his speech with loud applause.[17] Although Red Cloud did not achieve all his goals on the 1870 visit to the East, the journey could be considered a triumph in this respect—he had clearly won many friends and the support of the New York papers.

On 8 June, for example, the *New York Times* stated, "The clear conception which this unlettered savage possesses of what he claims as his rights, and what he is disposed to resent in his wrong, shows very plainly the necessity for treating with the leaders of the aboriginal 'nations' on some straightforward and intelligible principle. The attempt to cajole and bamboozle them, as if they were deficient in intelligence, ought to be abandoned, no less than the policy of hunting them down like wild beasts."[18] On 9 June, the *New York Herald* expressed admiration for Red Cloud's "logic and pathos," commenting, "Faithlessness on our part in the matter of treaties, and gross swindling by agents and their tools—the contractors—are at the bottom of all this Indian trouble."[19]

The Oglala delegation left New York City immediately after Red Cloud's speech and reached Pine Bluffs, Wyoming, on 24 June. They arrived at Fort Laramie on 26 June and found their village of about one thousand lodges at Rawhide Butte Creek.[20] There, according to a deprecatory story in the *Cheyenne Daily Leader*, Red Cloud returned to "the fleshpots of his own native camps and . . . his own dainty grasshopper omelet," after which the Oglalas went north, held a Sun Dance, and sent a war party against the Crows.[21] John Richard, Jr., got his wish, and the murder charge against him was withdrawn.[22]

While Red Cloud made an impression in Washington, the trip may have created an impression on him as well. Colonel Smith, the Oglala delegation's guide, remarked that Red Cloud left Washington with a sense that it would be futile for him to resist the government in the future, which was precisely the effect officials desired. Ironically, the sights Red Cloud had seen in Washington may have cost him some of his influence among the Sioux. Lieutenant William Quintin, a cavalry officer in Montana, wrote in May 1871 that he had heard about Sioux bands that had left Red Cloud to join Sitting Bull because they thought the things Red Cloud described from his Washington trip were not possible. Some Indians believed that the whites had been able to make Red Cloud see only what they wanted him to see.[23]

Soon after his return from Washington, Red Cloud traveled to the Powder River Country on a mission to persuade "non-treaty" Lakotas to make peace with the whites. He returned to Fort Laramie on 4 October, just in time for another meeting. The government had dispatched Felix R. Brunot and Robert Campbell to Fort Laramie, tasked with persuading the Oglalas to settle on a temporary agency at Rawhide Buttes, some forty miles to the north and away from settlers and emigrants. Brunot and Campbell served on the new Board of Indian Commissioners, a body created by Congress in 1869 to advise the Interior Department on Indian policy. As the board's chairman, Brunot had been present at some of Red Cloud's meetings earlier that year in Washington. Campbell, a veteran fur trader, had been a partner in the firm that founded Fort Laramie as a trading post in the 1830s. Red Cloud, Man Afraid of His Horses, and other Oglala notables met with Brunot and Campbell beginning on 5 October. Over several days, Brunot pressed the Oglala leaders to accept the Rawhide Buttes location with the threat of having their treaty goods made available only on the Missouri River if they refused. Red Cloud, as spokesman for the Oglala leadership, advocated for a trading post closer to Fort Laramie. The council ended on 8 October 1870 with no resolution of the matter.[24]

In 1871, government officials continued to press Red Cloud and the other Oglala leaders to settle at an agency. To this end, Felix Brunot, Colonel Smith, now commandant of Fort Laramie, and Joseph W. Wham met with the Oglala leadership at the fort on 12 June. Wham, a former army lieutenant from Illinois, had recently been appointed as agent for the Oglalas instead of Benjamin B. Mills, Red Cloud's choice for the job. The Episcopal Church had recommended Wham for the position, in accordance with President Grant's policy of inviting churches to nominate candidates for the Indian service.[25]

After opening prayers, Red Cloud stood and asked to speak first. The Great Spirit, he said, had raised both the white men and the Indians but had raised Indians first. The whites had crossed the "great waters" (the Atlantic Ocean) to reach the Indians' land. Red Cloud had given the whites land, and now he had "but a small spot of land" left, which he meant to keep. Further, he expected his people to be compensated for land they had already given up for railroad construction.[26]

While Red Cloud, Colonel Smith, and Felix Brunot exchanged many words on 12 and 13 June, the key issue was simple. The government representatives said that Red Cloud must select an agency site at once—one that was away from white settlements such as Fort Laramie—and that until this matter was settled, rations would be withheld. Red Cloud's response in open council was that he needed to talk to other Lakota leaders, including those who had not signed the Treaty of Fort Laramie in 1868, before taking any action. In private conversations with Brunot, Smith, and Wham, Red Cloud said he was willing to go north of the North Platte River to trade. However, he needed consensus among Lakota leaders, including twenty-nine Hunkpapa councilmen and twenty-six Oglala notables not present at Fort Laramie, before he could make such a commitment.[27] Red Cloud mentioned Black Twin and Charging Shield as examples of leading men that he should consult. "We will meet on the Cheyenne River," he told Smith and Brunot after the council session of 12 June. "I will get the men of sense, and will try and decide where to put the post."[28] On 13 June, Red Cloud and the government negotiators agreed that he would get fifteen days to make his consultations with Black Twin and the other absent Lakota leaders. If Red Cloud did not return within the allotted time with a decision on the agency's location, the officials would hold a council with the other chiefs present at the fort and resolve the issue without him. Brunot informed the other chiefs of this arrangement the next day, with Red Dog and American Horse indicating their approval.[29]

Red Cloud did not return to Fort Laramie within fifteen days. In July, after much negotiation, the remaining Oglala chiefs agreed to a temporary agency at a site then known as Little Moon, about one mile west of the Wyoming-Nebraska border near present-day Henry, Nebraska. The location was thirty-two miles downstream of Fort Laramie on the North Platte River—still outside the reservation boundaries specified in the 1868 treaty. Red Cloud did not visit the new agency until the next spring.[30]

On 21 March 1872, Red Cloud and his followers reached the agency on the North Platte River, where the chief met with Dr. James W. Daniels, Agent Wham's replacement. The compound at Little Moon was commonly referred to as the "Sod Agency" even though its major buildings were made of logs and a log stockade surrounded it. The facilities included two large storehouses, a blacksmith shop, corrals, stables, and employees' quarters. The agency was twenty-five miles from any appreciable timber, lacked good grazing for ponies, and was close to whites traveling the Oregon Trail, who created problems by selling the Indians liquor.[31]

At his 21 March meeting with Daniels, Red Cloud asked the agent to send his people's rations to Rawhide Creek near Fort Laramie, where Black Twin was camped. Daniels agreed to do so if the Oglalas would select a new site on the White River, thus removing them from the North Platte and the main transportation routes used by whites. At a council called to consider the matter on 10 April, chiefs Red Cloud and Little Wound asked the younger men to select the new site. They chose a location on the White River about a mile from the mouth of White Clay Creek near present-day Crawford, Nebraska.[32]

The decision of Red Cloud and Little Wound to defer to younger men in the matter of selecting a new agency location reflected the influence of warrior societies in Lakota affairs. The warriors rejected any agency site on ground suited for agriculture because they had no wish to become farmers—an occupation at odds with warrior status. Agent Daniels, reflecting his frustration with the chiefs' inability to impose the government's will on their people, wrote the commissioner of Indian affairs on 14 April suggesting that henceforth, agents should stop negotiating with councils and deal only with designated chiefs and headmen. Once again, the government's desire to identify paramount chiefs with whom it could conclude binding agreements clashed with the realities of Lakota politics.[33]

Red Cloud offered Daniels a solution to this political problem, suggesting that another trip to Washington would "show them the white man's ways," in the chief's words.[34] The agent quickly agreed, hoping that nontreaty Lakotas would follow Red Cloud's example and send representatives as well. On 17 May, a delegation that included Red Cloud, Little Wound, Red Dog, and twenty-four other Oglalas departed for Washington. Red Cloud and Red Dog were the only Oglalas in the group who had made the trip before. The Indians would deal with a new secretary of the interior, Columbus Delano, and a new commissioner of Indian affairs, Francis A. Walker.[35]

The Oglala delegation met with Delano and Walker at the Interior Department on 27 May. Red Cloud and Little Wound asked that the government supply their people with guns and ammunition for hunting—a request Red Cloud had made several times before without success. The chiefs also asked for a clothing issue. Delano promised that some ammunition would be provided to the treaty Oglalas. The next day, Secretary Delano said, the delegates could meet with President Grant, but the meeting with Grant on 28 May was a disappointment for the Oglalas. The president suggested that the tribe relocate to Indian Territory (present-day Oklahoma). Red Cloud responded that the Oglalas had agreed to the White River site. Grant remarked that the White River agency location was in Nebraska, not on the Great Sioux Reservation, and the Indians might have to move again in the future. Secretary Delano made a similar point the next day, telling the Oglalas that settlers would eventually want the land at White River and that they should consider moving south to Indian Territory. Nonetheless, Red Cloud and his colleagues remained committed to the White River location.[36]

In September, however, the Oglala council took a different position. Red Dog of the Ouyhpe band, Red Leaf of the Wazazas, and Blue Horse of the Loafers were willing to occupy a White River agency, but Red Cloud and Little Wound, under pressure from the young men of their respective bands, determined otherwise. In November, about four thousand Oglalas settled at White River, but Red Cloud and his band established winter quarters on Hat Creek, fifty miles to the west of the White River Agency, while part of Little Wound's band moved south to camp near Fort Mitchell, Nebraska. On 9 November, Red Cloud visited the Sod Agency and informed Daniels that he intended to return once a month for rations. The agent thwarted the chief's show of independence by sending the supplies to White River.[37]

In February 1873, Red Cloud told Daniels that he wanted his agency located at Hat Creek in one month. This location was less suitable for agriculture than the lands at White River, which may have been the point from the Oglalas' perspective. Finally, Edward P. Smith, yet another new commissioner of Indian affairs, sent a special committee chaired by Felix Brunot to deal with the matter. After a series of councils in May and June, those Oglalas not already at White River agreed to move to the new location. When loaded wagons were almost ready to leave the Sod Agency on 25 July, Red Cloud and a few Bad Face warriors made a last show of resistance. They demanded that the move be postponed until John J. Saville, who had been appointed to replace Agent Daniels, arrived. Red Cloud also renewed his requests for guns and ammunition, which he eventually received. Daniels, who remained on duty awaiting his successor's arrival, managed to persuade Red Cloud to let the move to White River begin on 27 July. By August, most Oglalas were on the move to what was then called Red Cloud Agency, where Indian Office personnel hoped their charges would adapt to Euro-American ways.[38]

Red Cloud's motivations in the matter of locating an Oglala agency can seem confusing on the surface. His initial reluctance to commit to an agency inside the Great Sioux

Reservation and his courting of Black Twin and other non-treaty Lakotas indicate a reassertion of old ways, while his eventual agreement with the government's wishes to end nomadic life suggests accommodation to a new order. This apparent contradiction was but one of many that Red Cloud exhibited as he sought to retain his power to lead, which was never absolute and always subject to sudden changes in the struggle between old and new and the increasing domination of Euro-American society. Historian Robert M. Utley aptly described Red Cloud's new negotiating style: "He gave in little by little and only after bedeviling his adversary through endless consultation with his fellow chiefs, irrelevant demands, bewildering changes of mind, qualified promises, theatrical bluster, and a host of niggling delaying tactics."[39] The next few years would be a time of unrest, resistance, and restriction, as some Lakotas chafed at reservation life while their non-treaty relatives in the "unceded" territory violently resisted aggressive white expansion. Although he did his best to protect his people's interests through negotiations, Red Cloud eschewed further warfare. As he said in a message to his northern Lakota relatives in the summer of 1872, "Listen to me, and [do] not go to war any more. You must carry on war yourself. I am done."[40]

8

Red Cloud Agency and the Black Hills 1873–1875

Red Cloud opened the ball by informing the Great Father that the whole white race were liars and he took occasion to remark that he made no exception of the company present, which included the President, the Secretary of the Interior, the Commissioner of Indian Affairs, and numerous subalterns and stipendiaries of the Departments. —Chicago Tribune, *21 May 1875*[1]

The Red Cloud Agency headquarters stood on a hill overlooking the White River. To the west was a high escarpment covered with pine trees, while rich buffalo grass grew on the adjacent prairies. The agency's buildings included a large warehouse, offices, the agent's house, a blacksmith shop, and some stables. A schoolhouse was added later. Two traders' stores were nearby. Red Cloud's camp was to the south of the agency, while other bands could be found in all directions. While the agency served primarily Red Cloud's people, it also distributed supplies for some Cheyenne and Arapaho groups. John J. Saville, a physician from Sioux City, Iowa, nominated by the Episcopal church, was the first agent.[2]

Saville's job was not easy. The agency experienced a good deal of chaos due to visits from so-called Northern Indians, aggressive bands that had not accepted the 1868 Treaty of Fort Laramie but still demanded food and other stores from the agent. Saville could not always refuse their demands with safety. At the same time, any supplies issued to non-treaty Indians reduced the quantity available for reservation Indians, who were already upset due to difficult living conditions and the government's attempts to repress their old ways. The unruly visitors harassed Saville and his

Grove Family Library 101 Ragged Edge Rd South Chambersburg, PA 17202
717-264-9663

Patron Receipt -
03/13/2017 05:13 PM

Red Cloud : Oglala legend / Jo
978.004 MC   4/10/2017

Thank You

employees, demanded arms and ammunition, threatened cowboys, and rode through the unfinished stockade shooting out windows. In addition, some of Little Wound's band left the reservation in the summer of 1873 to fight Pawnees on the Republican River, and some reservation Sioux joined their northern tribesmen in attacks on Northern Pacific Railroad survey parties in Dakota Territory. To cap it all, Indians at Red Cloud Agency belligerently resisted Saville's efforts to take a census, fearing that the count would result in reduced rations.[3]

Red Cloud, too, had his problems. His Oglala tribesmen had several roads open to them. Some, like Crazy Horse, refused to accept any of the 1868 treaty's restrictions and thus faced a potential response from the United States military. Others continued to roam free in the unceded territory defined in the treaty while refusing to move to the agencies. Still others became reservation Indians, who could choose either to cooperate with the agent or oppose his program to make them into farmers. As might be expected, the multiplicity of choices resulted in increased factionalism, splitting some bands between treaty and non-treaty groups. For example, the Bad Faces remaining in the Powder River Country now followed such leaders as Black Twin and Crazy Horse in Red Cloud's absence. Oglala bands were loose structures, and families faced no criticism for changing their affiliation from one band to another. The number of recognizable Oglala groupings increased over time, indicating the disintegration of tribal unity. Consequently, Red Cloud had to deal with more and more groups with diverging interests, limiting his ability to lead and his power to act. Because much of Red Cloud's band still remained in the Powder River Country, he had only a few lodges with him when he arrived at the new agency. Thus, only his core followers were on the White River to support him, but the government had named Red Cloud principal chief of the reservation Sioux and expected him to control all the reservation Oglalas. The Indian Office's inability to provide rations as promised and the principal freight contractor's failure to deliver annuity goods on time exacerbated the situation, as did the practice of guessing cattle weights at beef issue time and other irregularities.[4]

Agent Saville's 1875 annual report identified four distinct bands among the Lakotas living at Red Cloud Agency. The largest of these was called the true Oglalas or Old Smoke's people. Subdivisions of the true Oglalas included Man Afraid of His Horses's Hunkpatilas (Those Who Camp at the Horn), Red Cloud's Iteshichas (Bad Faces), and Yellow Bear's Tapishlechas (Spleens, also called Split Livers and Bents). These groups would soon operate separately. Although the elder Man Afraid of His Horses had signed the Treaty of Fort Laramie, he had no intention of living at an agency in 1868. However, as time went on, the buffalo became scarce and the people suffered. The old chief and his son, Young Man Afraid of His Horses, began to realize that the nomadic lifestyle could no longer be maintained. Young Man Afraid then used his position as a "shirt-wearer" and his family prestige to guide the Oglalas on a slow transition to a more fixed way of life on the reservation. As a result, Red Cloud often found Young Man Afraid's followers supporting the Indian agent's wishes rather than his own. The Man Afraid group came to be known as the Payabyas (Pushed Asides), possibly indicating its loss of influence due to Red Cloud's War in 1866–1868 and the government's recognition of Red Cloud as paramount chief afterward.[5]

The second band mentioned in Saville's report was the Kiyuksas or Cut–Offs (also translated as Breaks His Own, or Bit-the-Snake-in-Two). Little Wound led the group after his brother, the younger Bull Bear, died in 1865. Because Red Cloud had killed his father, the elder Bull Bear, in 1841, Little Wound considered Red Cloud an enemy but worked with his rival on occasions when their interests coincided. Black Bear, Pawnee Killer, and Whistler were three camp leaders among the Cut-Offs. In 1872, Whistler told the post commander at Fort McPherson, Nebraska, that the Cut-Offs desired to be one people to themselves, with an agent and reservation of their own.[6]

The Oyuhpes (Unloaded, Throw Down, Untidy), led by Red Dog, were the third band identified in Saville's report. Other prominent Oyuhpe headmen included Big Road, Slow Bull, High Wolf, and Big Foot. Red Dog lived to be a very old man and remained a cultural conservative to the end of his life. Consequently, Red Dog and Red Cloud were often allies.[7]

Finally, the fourth band named in Saville's report was the Wazhazhas, who were mostly Brules rather than Oglalas. Their leader, Red Leaf, often associated with the Oglalas at Red Cloud Agency from 1873 to 1877, but the group still considered themselves Brules. In 1877, Red Leaf's people transferred to the Spotted Tail Agency. After the Pine Ridge and Rosebud reservations were established in present-day South Dakota, the Wazhazha band split into groups led by Red Leaf and by Day, one settling at Pine Ridge and the other at Rosebud.[8]

Although Agent Saville did not mention the Wagluhe band (Loafers, In Breeders, Lives with Wife's Relatives, Sticks around the Fort) in his annual report, a contingent led by Blue Horse, the second son of Old Smoke, lived near Red Cloud Agency. Blue Horse prided himself on never having raised a hand to strike a white man and was the agent's stalwart supporter. After 1876, Blue Horse's influence as leader of the Wagluhes declined in favor of younger headmen such as American Horse and Three Bears. When the Oglalas settled at Pine Ridge in 1878, the Wagluhes split into three smaller bands led by Blue Horse, American Horse, and Three Bears respectively. According to George E.

Hyde, the number of recognized Oglala bands had grown to fourteen by 1883, indicating the success of the government's policy to break the power of the chiefs.[9]

The first incident of serious violence at the new Red Cloud Agency was the killing of agency clerk Frank Appleton on 9 February 1874 by a Minneconjou man, Kicking Bear, who had mistaken him for Agent Saville. Appleton was Saville's nephew. Unnerved by rumors that northern Sioux bands were preparing to fight the army, the agent had requested a military presence at Red Cloud Agency in January 1874. After Appleton's death, the agency's white employees became reluctant to stay at their posts without army protection, and Saville issued urgent calls for military assistance in concert with his colleague Edwin A. Howard, agent for Spotted Tail's people. The first troops arrived at Red Cloud Agency on 5 March and proceeded to establish a post called Camp Robinson (later Fort Robinson). The military encampment remained a powder keg through the summer.[10]

In October 1874, Lakota hostility to the army presence reached its apogee when disgruntled warriors chopped down the agency's flagpole and threatened to destroy both the agency and the military camp. Many younger warriors had opposed Agent Saville's plan to erect a flagpole because they considered it a sign of military domination. While Red Cloud knew of the plan to destroy the flagpole, he did nothing to stop it. The potential for conflict increased when troops from Camp Robinson arrived on the scene at Agent Saville's request, but Young Man Afraid of His Horses and other leaders intervened to prevent violence. The incident was an example of Red Cloud's deference to the unreconstructed element among the Lakotas.[11]

Further complicating the situation, some Lakotas continued to raid white settlements and attack Northern Pacific Railroad surveyors. Government officials believed that some raiders committed depredations and then fled back to the reservations through the Black Hills, including men from Spotted Tail's and Red Cloud's camps. Citing such violent incidents, Lieutenant General Philip H. Sheridan ordered Lieutenant Colonel George A. Custer to explore the Black Hills and find a suitable place for a military post in the spring of 1874.[12] Although not officially his principal objective, Custer would also investigate rumors that paying quantities of gold existed in the region. General Sheridan may have had mixed motives. Dr. Valentine T. McGillycuddy, who accompanied the expedition and later served as agent on the Pine Ridge reservation, wrote some fifty years later that Sheridan had admitted to being "under the pressure of the railroads to send Custer into the Hills in 1874 to find gold, and to start up a mining excitement that would force the miners into the Hills and split the reservation open in spite of and in violation of the Sioux Indian treaty of 1868."[13]

Custer's expedition was the first large Euro-American group to penetrate the Black Hills, then one of the last unexplored regions in the country. Rumors of gold in the area had been prevalent for decades. As early as 1804, fur trader Régis Loisel had told Spanish authorities in New Orleans about Black Hills gold. Dakota Territory's legislature sent memorials to Congress in 1862 and 1866 that claimed conclusive proof of great mineral wealth and requested a geological survey of the Black Hills. The claims received further promotion in 1869 when Ferdinand V. Hayden's descriptions of gold discoveries in streams near the Black Hills appeared in the *Proceedings of the American Philosophical Society*.[14]

The Black Hills expedition was an extravaganza, including ten companies of cavalry, two companies of infantry, six army engineers, two civilian miners, a photographer, a geologist, a topographer, a naturalist, civilian teamsters and herders, four newspaper reporters, a group of about one hundred Indian and two white scouts, and the Seventh Cavalry's regimental band. In all, Custer's party numbered about one thousand men, one hundred and ten wagons, and two thousand animals. The caravan left Fort Abraham

Lincoln, near Bismarck in present-day North Dakota, on 2 June and returned on 30 August. The expedition's confirmation of gold in the Black Hills became public even before the troops returned to Fort Abraham Lincoln.[15]

Reaction to Custer's report filled the country's newspapers and triggered a gold rush. Headlines in the *New York Tribune* on 10 August proclaimed a "New Gold Country," while the *New York World* declared, "The Reports of Surface Gold . . . Fully Confirmed" and "The Garden of America Discovered" on 16 August.[16] As the *Chicago Inter-Ocean* opined on 1 August, "There could be hardly a more fortunate event for the country. . . . the Black Hills will be opened and pull us through by providing thousands with occupations and stimulating trade and enterprise in every direction."[17] The *Bismarck Tribune* devoted two full columns to the story on 12 August, stating that the Hills would yield "Gold and Silver in Immense Quantities."[18]

Early resistance to white acquisition of the Black Hills among those less interested in profiteering was based upon the belief that Indians held the region to be sacred ground. For example, an April 1874 article in the *Omaha Weekly Bee* cited Joe Belton, a white man who had married one of Red Cloud's daughters and lived with the Oglalas for two years: "The Indians told him [Belton] that no paleface could ever enter there [the Black Hills]; it was the dwelling place of the departed spirits of the red men, preparatory to the final journey to the happy hunting grounds." Belton also told the newspaper that Indians often visited the Hills to commune with the spirits of the dead and jealously guarded the area from white intruders.[19]

On the other hand, Lieutenant Colonel Richard I. Dodge, who escorted Professor Walter P. Jenney to the Black Hills in 1875 to do a more thorough geological study, asserted that the area had little meaning for the Sioux. Dodge denied that the Black Hills had ever been "a permanent home for any Indians," stating that while small parties visited the Hills to cut lodge poles, "all the signs indicate that these were mere sojourns of the most temporary character."[20] Dodge reported that Indian informants had given him several reasons why they did not enter the Hills, including the presence of spirits and frequent strong storms. Lieutenant Colonel Dodge not only forwarded his assessment to his military superiors in 1875, but he also published the same information in book form the following year.[21] Dodge's writings supported federal officials' moves to acquire the Black Hills, while Lakotas, not possessing a written language, could not present a dissenting view to Euro-American readers.

Nevertheless, leaders of the reservation Oglalas and Brules, Red Cloud foremost among them, did not prize the Black Hills as much as some later historians suggested. As early as 2 January 1875, Red Cloud and Spotted Tail let it be known that they were interested in going to Washington with some of the most influential young chiefs in their respective bands to negotiate the sale of the Black Hills. In all their discussions concerning disposition of the Hills, Red Cloud and Spotted Tail never once argued that the land was sacred. That is not to say the two chiefs denied the region's mythical or supernatural significance—they did not—but they equated its sacred value with its economic value, using the language best suited to white ears. To Red Cloud, the power of *Wakan Tanka* was concentrated in all its multiplicity in the Black Hills, where it expressed its beneficence for the Lakotas. For example, in a September 1875 speech to government negotiators, Red Cloud said: "I know it well, and you can see it plain enough that God Almighty placed those hills there for my wealth."[22] In the end, the real motivation for the chiefs' willingness to consider the lease or sale of the Black Hills remains unclear. Perhaps Red Cloud saw the loss of the Black Hills as inevitable, but his acquiescence greatly diminished his influence among many of his contemporaries, and it remains one of the black marks on his reputation among Lakotas today.

Sitting Bull, as did others, opposed the sale for practical reasons. Standing Bear recalled that Sitting Bull likened the Black Hills to a "food pack" because of its resources. As Standing Bear later explained to John G. Neihardt, "Indians would rove all around, but when they were in need of something, they could just go in there [the Black Hills] and get it."[23] A newspaper reporter accompanying Custer's Black Hills expedition wrote that Indians familiar with the area mentioned its desirability as an agricultural and stock-raising country. Lieutenant James Calhoun, one of Custer's officers, wrote that Indian ponies wintering in the Black Hills all came out fat in the spring and that the land was well adapted for stock-raising operations. Thus, both Lakotas and whites knew the Black Hills contained valuable natural resources.[24]

As if to prepare himself for his new role as a warrior-diplomat exercising his skills against the whites' leaders, Red Cloud began to modify his dress and manner of living. To white ethnocentrists, he appeared to have adopted white ways. As the *Cheyenne Daily Leader* commented in April 1875, "We are informed that aside from living independently in a tepee, the noted chief apes the manners and customs of the white men to a great extent, wearing plain citizen's clothes, eating from a table set with china and furnished with quite a respectable array of well-cooked food, and driving back and forth in a light wagon."[25] A month later, a *New York Sun* correspondent's description of Red Cloud on the reservation displayed condescension but also a fear and grudging respect of the chief that reflected white nineteenth-century attitudes:

Red Cloud is very different in appearance and reputation from his more northern ally and rival [Spotted Tail]. I encountered him at the trader's store, clothed in a mongrel, nondescript suit; partly soldier, partly civilian, but above all Indian. To me, his countenance was most sinister and treacherous, and by the successful practice of the characteristic indicated in his visage, has won his present notoriety and eminence in his tribe. He is feared by both whites and Indians, but trusted by no one; and to me was so surly and uncommunicative, that I did not seek to prolong my interview.[26]

On 19 May 1875, Red Cloud, Spotted Tail, and other Lakota delegates met with President Ulysses S. Grant at the White House. The chiefs did not make a favorable impression on their host. When Red Cloud addressed the group, he informed the Great Father that the "whole white race were liars" and the present company was no exception.[27] Spotted Tail snapped his fingers in the president's face, saying that he did not want to meet with the secretary of the interior and the commissioner of Indian affairs, who had lied to him. Grant informed the chiefs, however, that he was not prepared to discuss matters, and the delegation withdrew.[28] On 26 May, the delegation met with President Grant in his private office. Governors John L. Pennington of Dakota Territory and John M. Thayer of Wyoming Territory, Commissioner of Indian Affairs Edward P. Smith, and Acting Secretary of the Interior Benjamin R. Cowen were also present. Coming quickly to the point, Grant urged the Indians to give up the Black Hills and threatened to starve them if they refused. In a sharp speech, Grant told the Indians that the country in which they lived was "entirely incapable of supporting them. . . . The food and provisions . . . which have been given to them for the last two years, has [sic] been a gratuity on the part of the Congress. These may be taken from them at any time without any violation of the treaty." Grant also pointed out that the white population of the West was growing so rapidly "that before many years it will be impossible to fix any point within the limit of our Territories where you can prevent them from going." Declaring that strife over the Black Hills might lead to war,

he menaced, "Should such trouble occur . . . it would necessarily lead to the withholding, for a time at least, of the supplies [the] government has been sending them. . . . I want
the Indians to think of what I have said."[29] Grant ended the
meeting with a promise that the Indians could have a written copy of his speech for their interpreters to read to them.
The *Cheyenne Daily Leader* reported that the delegation then
withdrew, "evidently disappointed in not having an opportunity to reply to the president."[30]

Red Cloud and his fellow Lakota delegates did not resolve the Black Hills issue during their 1875 trip to Washington. Accordingly, on 18 June, the secretary of the interior
appointed a commission chaired by Senator William B. Allison of Iowa, a close friend of President Grant's, to negotiate the purchase of the Black Hills from the Sioux or to
obtain mining rights there. The Allison Commission also
included Senator Timothy O. Howe of Wisconsin, the Rev.
Samuel D. Hinman, an Episcopal missionary to the Yankton
Sioux, and Geminen P. Beauvais, a trader who had served as
an interpreter for councils at Fort Laramie in 1866. John S.
Collins, post trader at Fort Laramie, was the commission's
secretary. The group planned a grand council near the Red
Cloud and Spotted Tail agencies for late summer. In the
meantime, Brigadier General George R. Crook, commanding the army's Department of the Platte, issued a proclamation on 29 July ordering miners and other unauthorized
whites to leave the Black Hills and the other unceded Indian
territories described in the Treaty of Fort Laramie before 15
August. Crook's proclamation directed these whites to stay
out of Indian territory until new arrangements were negotiated with the Indians.[31]

Several members of the Allison Commission arrived
at Red Cloud Agency on 4 September and met in private
for the next few days. In order to placate Spotted Tail, the
commissioners had initially agreed to hold the council on
Chadron Creek midway between Red Cloud Agency and

the Brules' headquarters at Spotted Tail Agency. However, with the support of Red Cloud and other headmen, officials changed the venue to a site on the White River eight miles from Red Cloud Agency, where the commissioners had found comfortable quarters. The council site was also near Camp Robinson, where two companies of cavalry were stationed.[32]

In a short time, thousands of Indians gathered near the council site while their ponies grazed nearby. Newspapers described the council as the greatest gathering of Indians ever assembled on the North American continent. According to the *Omaha Weekly Bee*, participants included Henry W. Bingham, the agent at Cheyenne River Agency, and six thousand members of the Sans Arc, Two Kettle, and Blackfeet bands. John Burke, agent at Standing Rock Agency on the Missouri, arrived with between three and four thousand Upper and Lower Yanktonais, Hunkpapa, and Blackfeet Sioux. Added to these numbers were an estimated fourteen thousand residents at Red Cloud Agency and eight thousand from Spotted Tail Agency. At the commissioners' request, Young Man Afraid of His Horses visited Sitting Bull's camp and returned with Little Big Man and five other representatives of the nontreaty Sioux. Thus, according to the *Bee*, the outlook promised "such a gathering of Indians to ratify a treaty as has never been collected at any previous time for any purpose."[33]

The commissioners took their time getting ready for the council, engendering some complaints from the Indians. In the meantime, Red Cloud and Spotted Trail jockeyed for position. For example, the Brule chief sent couriers to welcome the Missouri River Sioux on their approach to Spotted Tail Agency, then gave them a big feast and a gift of one hundred fifty ponies upon their arrival at Chadron Creek. Not to be outdone, Red Cloud hosted his own feast and gave the new arrivals one hundred eighty ponies, leading the *Omaha Bee*'s reporter to compliment the chief's shrewd diplomacy.[34]

Like the commissioners, the Sioux chiefs did not immediately press for a substantive meeting, with some absent due to illness or because their people were busy receiving their allotted rations. The commissioners waited, chatted, received some Indian visitors for short discussions, and adjourned. When Brigadier General Alfred H. Terry, a commissioner and the commander of the military Department of Dakota, finally arrived on 15 September, large delegations of Indians paid their respects to the commission. The council convened for substantive business on 20 September, but Red Cloud did not attend that day's session. Reportedly, he was dissatisfied with the meeting location.[35] Trouble appeared likely on 23 September, when Little Big Man's armed followers took a position behind the cavalry guard from Camp Robinson. Riding naked on an iron-gray horse and brandishing his weapons, Little Big Man charged toward the council tents, shouting in his native tongue: "My heart is bad. I have come for white men's scalps."[36] He threatened to kill anyone who would sell his land. The officer commanding the commission's military escort prepared for the worst, but Young Man Afraid of His Horses intervened to salvage the situation. Young Man Afraid sent his followers to check the onrush of Little Big Man's warriors and stationed guards to the soldiers' rear. At the commissioners' request, Red Cloud and Spotted Tail each chose four men to help maintain order.[37]

Heated discussions continued for three days in the Indian camps. Little Big Man's outburst reflected a serious division of opinion between those Sioux unwilling to give up the Black Hills under any circumstances and those, like Red Cloud and Spotted Tail, who might consent to sell if the price were high enough. Showing frustration, the commissioners called twenty leading chiefs to their quarters at Red Cloud Agency on 26 September, telling the chiefs that they must come to an agreement among themselves in time for a meeting the next day.[38] The chiefs, however, had made up their minds, and Red Cloud delivered the following speech:

> There have been six nations raised, and I am the seventh and I want seven generations ahead to be fed. . . .
> These hills out here to the northwest we look upon as
> the head chief of the land. My intention was that my
> children should depend on these hills for the future.
> I hoped that we should live that way always hereafter.
> That was my intention. I sit here under the treaty which
> was to extend for thirty years. I want to put the money
> that we get for the Black Hills at interest among the
> whites, to buy with the interest wagons and cattle. We
> have much small game yet that we can depend on for
> the future, only I want the Great Father to buy guns
> and ammunition with the interest so we can shoot the
> game. For seven generations to come I want our Great
> Father to give us Texan steers for our meat. I want the
> Government to issue for me hereafter, flour and coffee,
> and sugar and tea, and bacon, the very best kind, and
> cracked corn and beans, and rice and dried apples and
> saleratus and tobacco, and soap and salt, and pepper,
> for the old people. I want a wagon, a light wagon with
> a span of horses, and six yoke of cattle for my people. I
> want a sow and a boar, and a cow and a bull and a sheep
> and a ram, and a hen and a cock, for each family. I am
> an Indian but you try to make a white man out of me. I
> want some white men's houses at this agency to be built
> for the Indians. I have been into white people's houses,
> and I have seen nice black bedsteads and chairs and I
> want that kind of furniture given to my people. . . .
> Maybe you white people think that I ask for too much
> from the Government, but I think those hills extend
> clear to the sky—maybe they go above the sky, and that
> is the reason I ask so much. I think the Black Hills are
> worth more than all the wild beasts and all the tame

beasts in the possession of the white people. I know it well, and you can see it plain enough that God Almighty placed those hills there for my wealth, but now you want to take them from me and make me poor, so I ask so much that I won't be poor."[39]

Red Cloud had delivered the Lakota platform. His use of the concept of interest showed a new and sophisticated awareness of the white man's ways. The demand for livestock revealed the central role the tribe intended cattle to have in its future economy and a willingness among some Sioux to live as ranchers rather than crop farmers, a possibility that future government policy would effectively deny them.[40]

Red Cloud's speech, in concert with the words of other Indian spokesmen making similar demands, shocked the Allison Commission. Nevertheless, the commissioners submitted a proposal to lease the Black Hills for $400,000 per annum or buy them for $6 million, payable in fifteen annual installments. The commission also offered to buy the Bighorn country for $500,000 over a ten-year period. Spotted Tail and Red Cloud expressed surprise at the government's low price for the Black Hills, saying that they would have to refer the matter to an Indian council and thus could give no answer for two days. According to one eyewitness, the final Indian demands for the Black Hills included six yoke of oxen with wagons, one span of horses with a wagon, and other livestock and furniture for each family. Tribes were to receive subsistence annuities for six hundred years and would control the appointment of all Indian agents and reservation employees. Troops were to leave the reservation, and Christian religious instruction was to be provided by Catholic missionaries instead of the Episcopal Church. Furthermore, only one road—the route from Bismarck— would be permitted to enter the Black Hills, and the Great Sioux Reservation was to be extended to the Platte River in Nebraska. The chiefs refused to consider parting with their lands on the Powder and Bighorn Rivers, demonstrating resolve by putting their hands over their ears at the mention of such a sale. The astounded commissioners later claimed that the chiefs demanded $70 million for the Black Hills at one point. In the end, the Allison Commission concluded that a Black Hills purchase agreement was impossible for the time being, blaming the northern Sioux for the impasse. Red Cloud was absent from the commission's final session with the Indians, choosing instead to attend an Arapaho Sun Dance a few miles away.[41]

After returning to Washington, the commission prepared its final report, which recommended that Congress determine a fair price for the Black Hills and present it to the Indians as a final offer. If it was rejected, the government should end subsistence allotments. The report further recommended that the Sioux then residing in northwestern Nebraska be forced to occupy lands within the Great Sioux Reservation in present-day South Dakota and called for army officers (rather than civilian agents) to account for and issue supplies to the tribes. In a radical social move, the commissioners recommended that all Sioux children between the ages of six and sixteen be separated from the adult population and sent away to industrial schools. Male Sioux were to work for their subsistence, with the amount awarded to be based on performance.[42]

The Allison Commission's failure represented a major setback for miners and other Euro-American expansionists. In perspective, it was one of the most catalytic and divisive incongruities in the history of South Dakota and the Northern Plains in the last half of the nineteenth century. Had the commission succeeded, the Great Sioux War of 1876–1877 might have been avoided, and the Lakota people's resentment that continues to this day might have been mitigated. The lost opportunity affected generations to come.

During this same time frame, Red Cloud had reaped the fruit of one of his more effective negotiating techniques.

In November 1874, Professor Othneil Charles Marsh, a Yale vertebrate paleontologist, had visited Red Cloud while hunting fossils in the Badlands. The chief gave him samples of bad food and tobacco and filled him with stories of corruption on the reservation. To obtain Red Cloud's support for his explorations, Marsh agreed to air this information with the commissioner of Indian affairs and the public. After an unsatisfactory meeting with Commissioner Edward P. Smith in March 1875, Marsh contacted President Ulysses S. Grant regarding his allegations the following month. Marsh's actions prompted an investigation into the affairs of Agent James J. Saville and annuity contractors at Red Cloud Agency. Informed of the professor's charges while in Washington with Red Cloud, Spotted Tail, and other Lakota delegates, Saville wrote Commissioner Smith on 5 June to declare his innocence. Saville's letter also announced his intention to resign once his name had been cleared.[43]

Headed by ex-governor Thomas C. Fletcher of Missouri, the committee investigating Professor Marsh's accusations went west in late July 1875. No samples of rotten flour or other food were found, but one fraudulent activity uncovered involved beef contracts. Marsh had claimed that Indians at Red Cloud Agency were being cheated concerning the number of pounds of meat received. The contractors and inspectors simply guessed at the weights of cattle by looking at the animals. For example, according to the *Omaha Daily Bee*, beef contractor James W. Bosler had purchased twelve or thirteen thousand cattle, which had been driven straight through from Texas. They averaged not more than six hundred fifty pounds a head but were sworn to weigh up to twelve hundred pounds each. As a result of the investigation triggered by Marsh's allegations, the government adopted a system of weighing cattle on scales at all Indian agencies. Red Cloud's charges concerning spoiled pork were refuted, but it was found that the contractor, exploiting a clerical

Red Cloud posed for a portrait by Alexander Gardner in 1872 while on his second visit to Washington, D.C. South Dakota State Historical Society

FRANK LESLIE'S
ILLUSTRATED
NEWSPAPER

Entered, according to the Act of Congress, in the year 1870, by FRANK LESLIE, in the Clerk's Office of the District Court for the Southern District of New York.

No. 770—VOL. XXX.]   **NEW YORK, JULY 2, 1870.**   [PRICE 10 CENTS. $4.00 YEARLY. 13 WEEKS, $1.00.

MUSIC FOR THE MILLION.

"Fire low. Wait till you see the white of their eyes. Aim at their waistbands." These are the orders attributed to General Putnam to his command at Bunker Hill. They express the general idea, that people are apt to aim too high, that they hit only some extraordinary seven-footer—and not him, if by chance he should be tying his shoe, or in any but an erect position.

The habit of firing over people's heads is not an uncommon one to-day, and in very many ways. We see it in preachers in the pulpit, lawyers before a jury, less commonly but yet sometimes in stump-speakers and political ora-

Indian Delegation.   Peter Cooper Esq.   Interpreter.   Red Cloud.   Indian Warriors.
Red Dog.   Rev. Dr. Crosby.

NEW YORK CITY.—THE SIOUX CHIEF, RED CLOUD, IN THE GREAT HALL OF THE COOPER INSTITUTE, SURROUNDED BY THE INDIAN DELEGATION OF BRAVES AND SQUAWS, ADDRESSING A NEW YORK AUDIENCE ON THE WRONGS DONE TO HIS PEOPLE.—FROM A SKETCH BY OUR SPECIAL ARTIST.—SEE PAGE 247.

Frank Leslie's Illustrated Newspaper *featured Red Cloud addressing an audience at Cooper Union in New York City on the front page of its 2 July 1870 issue.* South Dakota State Historical Society

Garrick Mallery recorded this pictograph representing Red Cloud's name for the Smithsonian Institution's Bureau of American Ethnology in the 1880s. Mallery, *Picture-Writing of the American Indians* (New York: Dover Publications, 1972)

Red Cloud appears here with an Oglala delegation to Washington, D.C., probably in 1875. Seated, from left, are Red Dog, Little Wound, Red Cloud, American Horse, and Red Shirt. The man standing is believed to be interpreter John H. Bridgeman. Denver Public Library, Western History Collection, call no. Z-2298

This well-known portrait of the Oglala chief was taken by Washington, D.C., photographer Charles M. Bell in 1880. Library of Congress

Red Cloud and his wife Pretty Owl stand outside their house near Pine Ridge reservation headquarters around 1890. Their home was reportedly the only two-story house on the reservation. The man at left is unidentified. South Dakota State Historical Society

The woman in this interior view of Red Cloud's house at Pine Ridge is believed to be Pretty Owl, the chief's wife. Clarence G. Morledge photographed the scene in 1890 or 1891.

Red Cloud (left) attended a performance of Buffalo Bill's Wild West at Madison Square Garden in New York City in 1897. Following the show, he posed with William F. ("Buffalo Bill") Cody and fellow Oglala chief American Horse for this photograph by D. F. Barry. South Dakota State Historical Society error in the contract, had provided the lowest grade of pork possible.[44]

In its October 1875 final report, the committee declared that Agent Saville was "incompetent and unfit for the position" he occupied and that he should be discharged "without delay," citing "a nervous and irritable temperament, inordinate loquacity, undignified bearing and manners, a want of coolness and collectedness of mind, and of firmness and decision of character." The committee also determined, however, that there was "no proof to sustain" the assertion that Saville had conspired with contractors to commit fraud.[45] The Indian Office accepted Saville's resignation in the wake of the committee report. His replacement, James S. Hastings, took charge of the Red Cloud Agency on 3 December 1875.[46] In this instance, Red Cloud's technique of persuading men from various professions and parts of the country to carry his messages to government officials and politicians had proven beneficial.

9

The Great Sioux War and Its Results 1876–1877

We took away their country and their means of support, broke up their mode of living, their habits of life, introduced disease and decay among them, and . . . against this they made war. Could anyone expect less?
—General William T. Sherman[1]

As large numbers of miners surreptitiously entered the Black Hills and tension between the United States and the Sioux tribes increased, President Ulysses S. Grant held a meeting at the White House on 3 November 1875 to discuss the situation. Attendees included Secretary of the Interior Zachariah Chandler, Commissioner of Indian Affairs Edward P. Smith, Secretary of War William W. Belknap, Lieutenant General Philip H. Sheridan, and Brigadier General George R. Crook. The meeting produced two decisions. First, although non-Indians would officially remain barred from the Black Hills, the army would no longer enforce the ban. Second, the non-treaty Sioux were to leave the unceded Indian territory defined in the 1868 Treaty of Fort Laramie and move to reservations. In accord with the plan, on 6 December 1875, the commissioner of Indian affairs ordered the agents for the Sioux and Cheyenne tribes to inform the roving bands and non-treaty dissidents that they must go to their reservations by 31 January 1876 or be considered hostile. Runners soon carried the message to the Indians' winter camps. When few Indians appeared at the agencies by the deadline, the commissioner turned the matter over to the army. General Sheridan then took action, beginning what is probably the best-known and most-controversial campaign of the Plains Indian wars.[2]

As the army readied for its task, reservation Indians were suffering from hunger. On 24 April 1876, Captain William H.

Jordan, commanding Camp Robinson, wrote Lieutenant Colonel Luther P. Bradley at Fort Laramie to report that the Indians at Red Cloud Agency had been on the verge of starvation for days. Chiefs Red Cloud and Pawnee Killer had begged Captain Jordan for food to feed their families. In Jordan's opinion, the government's failure to feed the reservation Indians adequately came at the worst time. If beef did not arrive soon, Jordan predicted, reservation Indians would be forced to "join the hostiles to keep from starving."[3] James S. Hastings, the civilian agent at nearby Red Cloud Agency, also complained of inadequate supplies to his superiors at the Office of Indian Affairs. Undoubtedly, the scarcity of food at the agencies led many more Sioux to head north, where some of them would take part in the hostilities then underway.[4]

General Sheridan's plan for the campaign against the non-treaty Sioux and Cheyennes relied upon three converging columns. Brigadier General Alfred H. Terry's force was to move west from Fort Abraham Lincoln near present-day Bismarck, North Dakota, while Colonel John Gibbon would march east with troops from Forts Shaw and Ellis in Montana Territory. Brigadier General George Crook's troops were to move north from Fort Laramie and Fort Fetterman. General Sheridan believed that each force was strong enough to defeat any Indians encountered. There was no dictum to meet at a certain place at a certain time. Rather, the plan was to force the Indians into a general area where they could be engaged by any of the columns.[5]

First into the field was Crook's force of nine hundred men, which moved north from Fort Fetterman on 1 March 1876. The column's initial encounter with Indians occurred on 17 March near present-day Broadus, Montana. Troops led by Colonel Joseph J. Reynolds destroyed a Cheyenne village, but the Cheyennes' counterattack forced a hasty withdrawal. The Indians also recaptured their pony herd from Reynolds's men. Unhappy with his subordinate's performance, Crook recommended that Colonel Reynolds be court-martialed. Crook also reported to General Sheridan that evidence showed "that these Indians are in co-partnership with those at the Red Cloud and Spotted Tail agencies, and that the proceeds of their raids upon the settlements had been taken to these agencies and supplies brought out in return."[6]

Crook again took the field in late May with some fourteen hundred men. After an indecisive skirmish with the Sioux on 9 June, he moved his command to the headwaters of Rosebud Creek. On the morning of 17 June, the one hundredth anniversary of the Battle of Bunker Hill, he met a combined Sioux and Cheyenne force under Crazy Horse, Two Moon, and other war chiefs, probably numbering six hundred to one thousand warriors. The all-day battle that followed ended with few casualties on either side. Crook withdrew the next day to his base camp on Goose Creek.[7]

General Terry's command left Fort Abraham Lincoln on 17 May and united with Colonel Gibbon's column by 9 June. On 21 June, the commanders made their final plans aboard the steamer *Far West* on the Yellowstone River. Lieutenant Colonel George A. Custer's cavalry force was to approach the Little Bighorn Valley, believed to hold a large Indian village, in order to block any retreat to the south. Colonel Gibbon, with whom General Terry traveled, was to ascend the Bighorn River with the slower infantry and enter the valley from the north. If all went according to plan, this joint offensive would occur on 26 June, but Terry gave Custer the freedom to act with dispatch if conditions warranted. Believing he had been discovered, Custer attacked the large village on 25 June. Overwhelming numbers of Sioux, Cheyennes, and Arapahos wiped out the five companies of the Seventh Cavalry Regiment under Custer's immediate command. However, General Crook's victory at Slim Buttes in September and his destruction of Dull Knife's village in November, coupled with Colonel Nelson A. Miles's defeat of

Sitting Bull and Crazy Horse in battles in the fall and winter of 1876–1877, spelled the end of freedom for the remainder of the Sioux and Cheyennes. On 22 July 1876, a month after Custer's defeat, the Interior Department conceded the supervision of all the Sioux reservations to the military at General Sheridan's request. Army officers became acting agents at the Red Cloud and Spotted Tail agencies.[8]

Red Cloud was not present on any battlefield during the Great Sioux War of 1876–1877. About fifty-five years of age when the war broke out, he no longer expected to participate physically in battle. He had led the reservation Oglalas and some Brules at Red Cloud Agency for eight years, but his influence among the Sioux who stayed in unceded territory had long since been eclipsed by other leaders, such as Big Road and the rising star, Crazy Horse. Moreover, Red Cloud repeatedly stated that he had signed the 1868 Treaty of Fort Laramie and had never broken it, and he had everything to lose by leaving the reservation to become a secondary figure among the nontreaty Sioux.

Red Cloud, however, was involved in the Great Sioux War in several ways. For example, he had a personal connection to the struggle through relatives and friends who had chosen to fight. As active hostilities broke out in the spring of 1876, Red Cloud reportedly lost a son-in-law who participated in Lakota raids against miners traveling to the Black Hills. Jack Red Cloud, the chief's son, also joined the non-treaty Sioux and fought in the Battle of the Little Bighorn.[9]

Even though he himself stayed on the reservation, Red Cloud refused to cooperate with military leaders, and he eventually drew sanctions because of it. When General Crook arrived at Red Cloud Agency on 14 May 1876 seeking to recruit Indian scouts for his next campaign, Red Cloud defiantly refused to participate. "The Gray Fox must understand," the chief declared, "that the Dakotas and especially the Oglalas have many warriors, many guns and ponies. They are brave and ready to fight for their country. They are not afraid of the soldiers or of their chief. Many braves are ready to meet them. Every lodge will send its young men, and they will say of the Great Father's dogs, 'Let them come!'"[10] According to a story published about thirty years later in the *Indian School Journal*, Red Cloud sent a band of warriors to waylay and kill General Crook when he left the agency, but the war party abandoned the task when they found that Crook had a company of cavalry escorting him.[11] Whether the alleged assassination plot was real or not, Crook suspected that Red Cloud secretly provided arms and supplies to those Sioux who fought the army. And, of course, reservation Indians left by the hundreds to join their kinsmen in unceded Indian territory in the spring of 1876 without interference from their chief. In effect, Red Cloud supported the Sioux war effort, but he did so in a way that protected his position among his immediate followers and with the government.

General Crook was not the only officer to suspect that Red Cloud's sympathies lay with his kinsmen who were at war with the United States. Writing to General Sheridan from Camp Robinson on 5 June 1876, Lieutenant Colonel Wesley Merritt reported:

I have seen the Indian agent and talked with Captain Jordan. It is thought that from 1,500 to 2,000 Indians have left the reservation since [the] tenth of May; a large proportion . . . are warriors. The agent is inclined to underestimate those who had gone. . . . [The agent] admitted reluctantly that Red Cloud had informed him that some of his and other principal['s] families had gone, but that they were absent to recover stock stolen by northern Indians. The Indians here are not friendly. . . . In fact they are generally hostile.[12]

Even in the midst of these increasing hostilities, miners continued to enter the Black Hills. In September 1876, Seth

Bullock, a new arrival himself, wrote that the Deadwood Gulch area in present-day South Dakota was "overdone," in the sense that too many gold seekers had come there too fast, and that a great many migrants remained idle and broke. Bullock also noted that Indians kept men from leaving the gulch so that prospecting in other areas was at a minimum.[13] On 5 September, the *Butte* (Mont.) *Miner* reported that Sioux warriors had killed eleven men near Rapid City and Crook City between 18 and 21 August, picking them off one by one.[14]

A major consequence of the Great Sioux War was that the government forced an agreement upon the agency Sioux to reduce the 1868 treaty reservation significantly, resulting in the loss of the Black Hills. William Vandevere, an inspector for the Office of Indian Affairs, visited Red Cloud Agency in late June 1876. After interviews with Red Cloud, Spotted Tail, Swift Bear, Red Dog, Fast Bear, and other leading chiefs, Vandevere reported, "one and all expressed a desire to live in reservations," and that they were perfectly willing to cede the Black Hills to the government on moderate terms.[15] Before sending another commission to the Sioux reservations, Congress, largely due to the influence of Nebraska senator Algernon S. Paddock, attached a condition to the Indian appropriation bill that became law on 15 August 1876. The proviso stated that no money would be spent for the support of the Sioux nation west of the Missouri River until the tribes had agreed to sell the Black Hills, surrender all Nebraska soil they occupied, accept their rations on the Missouri River, and move to the reservation in present-day South Dakota.[16]

On 24 August 1876, President Grant duly appointed a commission to present the government's new demands to the Sioux. Headed by George W. Manypenny, a former commissioner of Indian affairs, the commission held its first council session at Red Cloud Agency on 7 September. Manypenny and his colleagues delivered the government's terms to about one hundred fifty chiefs and headmen including Red Cloud, Red Dog, Man Afraid of His Horses, his son Young Man Afraid of His Horses, and the Oglala chief Sitting Bull (not to be confused with the Hunkpapa medicine man of the same name). The commissioners explained that they did not have the authority to change the agreement as written. (Technically, the agreement was not a treaty because Congress had passed legislation in 1871 to change the form of negotiation with tribes in order to give the House of Representatives more say in the making of Indian policy. Heretofore, the Senate had dealt with treaty-making without the House because only the Senate could ratify a treaty.) The Indians requested a written statement of the proposal and retired to their camp to consider the matter.[17]

Once they understood the proposed agreement's specific provisions, the chiefs at Red Cloud Agency and elsewhere generally opposed it. Under Article I, the Sioux would relinquish not only the Black Hills, some 7.3 million acres, but also their rights to the unceded hunting grounds in the Bighorn River basin as defined in Article XVI of the 1868 Treaty of Fort Laramie. The compensation offered for all of these concessions was a paltry nine hundred thousand acres of additional grazing land added to the north side of the Great Sioux Reservation. Furthermore, the proposal required the Sioux to agree to three roads across the reservation and to go to the Missouri River for their annuities.[18]

Article IV of the proposed agreement was of particular concern for the Lakotas because it threatened to take them far away from their home region. The article stated the president's belief that the only country where the Oglalas and Brules could hope for permanent improvement was Indian Territory (present-day Oklahoma) and directed that a Sioux delegation visit the area in the hope that the tribes might eventually abandon their treaty lands in exchange for better farmland to the south. The commissioners explained this provision by declaring that the Indians must depend mainly upon the cultivation of the soil for their future support and that their present reservation was quite unfit for the purpose. The superior climate and soil in Indian Territory and the promise that it was forever secure to the Indian people would, in the commission's opinion, be strong inducements for them to accept the agreement. The commissioners pointed out that the Lakotas could not live on their present reservation without government help, and the government was under no treaty obligation to continue food rations. If they declined to move, the commissioners stated, the Lakotas would have to go to the Missouri River to receive their annuity goods.[19]

Red Cloud and several other Oglala chiefs eventually signed the agreement on 20 September. Subdued by threats of no food and exile to Indian Territory, the chiefs and headmen made their marks on the paper but made no secret of their displeasure in doing so. Chief Fire Thunder signed with a blanket over his eyes, while the *New York Herald*'s reporter described Red Cloud as "grimly meditative," displaying "a sullen discontent" at the ceremony.[20] In all the discussions, no mention was made of the sacredness of the Black Hills.

Red Cloud made a speech detailing his people's needs:

I am a friend of the President and you men who have come here to see me are chief men and men of influence. You have come here with the words of the Great Father; therefore, because I am his friend I have said yes to what he has said to me, and I suppose that makes you happy. I don't like it that we have a soldier here to give us food; it makes our children's hearts go back and forth. I wish to have Major Howard [Edwin A. Howard, former agent at Spotted Tail] for my agent, and I want to have you send word to Washington, so he can come here very soon. If my young men come back and say that the country [Indian Territory] is bad, it will not be possible for me to go there. As for the Missouri river country I think if my people should move there to live they would all be destroyed. There are a great many bad men there, and bad whiskey; therefore I don't want to go there.[21]

Even though Red Cloud signed the agreement, his remarks at the signing ceremony indicate that he did so with misgivings. Nevertheless, he addressed the Brules at Spotted Tail Agency a few days later to recommend that they also accept the proposal. In response, Spotted Tail said that if his friends at Red Cloud Agency had refused to sign he would have held out with them, but because they had signed, he would do so as well.[22]

The agreement promised reservation Indians a daily individual ration of one-and-one-half pounds of beef or one-half pound of bacon, one-half pound of flour, and one-half pound of corn. For every one hundred individual rations, four pounds of coffee, eight pounds of sugar, and three pounds of beans or other equivalent foodstuffs were to be issued. These supplies were promised only until the government determined that the Indians were self-sufficient. Furthermore, the agreement specified that when the government built schools on the reservation, attendance would be mandatory in order for the children to receive their rations.[23]

Ignoring the 1868 treaty provision that required the approval of three-fourths of the adult males living on the Sioux reservation for any land cession, the commission merely obtained enough signatures at the various Sioux agencies in September and October 1876 to satisfy Congress. On 22 December, President Grant sent the agreement to the House and Senate for action. In the meantime, a delegation of young men led by Spotted Tail visited Indian Territory to investigate its desirability for future relocation. Shortly after their visit, Congress, pressured by the railroads, amended the document to prohibit explicitly any Sioux migration to Indian Territory until authorized by further legislation. Congress approved the amended agreement on 28 February 1877.[24]

Even though Red Cloud had signed the document ceding the Black Hills, General Crook still suspected the chief of providing covert assistance to the non-treaty Sioux who were fighting the army off the reservation. Crook took what he considered appropriate action. On 20 October 1876, he ordered Colonel Ranald S. Mackenzie to disarm and dismount Red Cloud, Red Leaf, and their followers. Mackenzie's troops surrounded Red Cloud's camp, seized over seven hundred horses, and took the Indians' weapons. This action would be a sore point in the reservation Oglalas' relations with the government for many years. Red Cloud spent the night in the guardhouse at Camp Robinson and later told a reporter that as he was marched past the flagstaff his heart was heavy and that it came into his mind to break away from his guards, run under the flag, and end his life by his own hand. The chief, however, decided against the action as night came on. Crook delivered another blow to Red Cloud on 24 October, deposing him as the chief of the agency Indians and appointing Spotted Tail of the Brules to be overall chief of the Sioux.[25]

Red Cloud soon regained his composure. When asked later if he felt bad about Spotted Tail being made chief over his band, he replied that Spotted Tail could never be a true chief over his Oglala people. He was still their leader, and they would obey him.[26] Interpreter William Garnett, the son of an Oglala mother, agreed with Red Cloud's assessment. In 1907, Garnett told Eli Ricker: "At the council which was held at Red Cloud Agency [24 October], General Crook deposed Chief Red Cloud and made Spotted Tail the chief over the Indians. . . . So far as Red Cloud was interested, this little affected him. Much less affected [were those] who acknowledged his chieftainship, which was not denied in any quarter among them."[27]

In an 1883 letter to fellow military officer Oliver O. Howard, Crook explained his motivation:

> One word in reference to Red Cloud himself: He did not take an active part in the troubles of 1876–77 and I can accuse him of no overt act of hostility, but in every way in which he could [he] manifested his sympathy for the Indians on the war path, sent them supplies of ammunition, aided and assisted them with information in regard to our movements, he showed himself to be an alert, ill disposed and dangerous rascal. When enough testimony had accumulated . . . I determined to lose not a moment in stripping him of every vestige of authority. My action wasn't taken a moment too soon. Red Cloud and Red Leaf were overtaken a considerable distance away from the Agency, on their way to the north to join the hostiles, as Spotted Tail and other Indians believed and averred. I deposed Red Cloud and put Spotted Tail in command of all the Sioux.[28]

As the conflict between the warring Plains Indians and the United States Army began to wind down—with Crook breaking the power of the Northern Cheyennes in the Battle of the Red Forks of Powder River in late November 1876 and Miles besting Crazy Horse at Wolf Mountains in January 1877—Red Cloud made a number of compromises to regain his standing with the whites. These included symbolic as well as substantive acts. The first was enlistment as an army scout. Lieutenant William P. Clark, an aide to General Crook, had the authority to recruit reservation Indians, as well as Indians who came in to surrender, for this duty. Indian scouts were issued firearms and could keep horses, which made enlistment popular on the reservation. Red Cloud agreed to service with the rank of sergeant. Writing to Lieutenant John Bourke, another of Crook's aides, on 28 March 1877, Clark explained why he had enlisted Red

Cloud: "I am quite convinced it was a good thing to take Red Cloud. . . . He has felt keenly and bitterly his downfall and he will work hard and earnestly to regain something of his former prestige and power. Not for the love he bears us but he is keen and shrewd enough to know that it is the only way that his wounded pride may be healed and part of his ambition gratified. I was convinced he could be used to advantage."[29]

The chief also agreed to help induce Crazy Horse to surrender. In mid-April 1877, Red Cloud set out with a delegation of eighty Indians to persuade the famous warrior to come in. Crazy Horse did surrender, but his presence turned out to be a problem at Red Cloud Agency, where he also agreed to enlist as a sergeant of Indian scouts. Moreover, Crazy Horse had his own substantial following among the Oglalas, which, according to one observer, caused Red Cloud, Spotted Tail, and their respective supporters to become jealous of him.[30]

Intrigue, misunderstanding, outright lying, and vindictiveness led to Crazy Horse's arrest on 5 September 1877. Taken to Camp Robinson and told that he was going to see the post commander, he found himself inside the guardhouse door. Attempting to break away from his escort of Indians and military officers, he grabbed his knife, slashed the arm of Little Big Man—who was trying to hold him back—and backed out the door, where one of the guards bayoneted him in the side. Carried next door to the post adjutant's office, Crazy Horse was attended by the post physician, Dr. Valentine T. McGillycuddy, who later served as agent on the Pine Ridge reservation. The doctor soon realized that Crazy Horse was mortally wounded and administered morphine to ease his pain.[31] As McGillycuddy later related the story to a friend: "[With me were] old man Crazy Horse [the chief's father]; and chief Touch the Cloud (Mahpia Yutan), six foot four in height. When Crazy Horse died, this chief drew the blanket over the face of the dead man and standing up, pointed to the body and said: 'There lies his lodge.' Then [he] pointed up [and said], 'The chief has gone above.'"[32]

Red Cloud's role in the death of Crazy Horse is clear. Along with other agency chiefs, Red Cloud had attended a meeting with General Crook on 3 September 1877 to consider what to do about Crazy Horse. Crook proposed to send a party of reservation Oglalas to kill the war chief in his camp, to which Red Cloud agreed. When circumstances foiled Crook's plan, Lieutenant Colonel Luther P. Bradley, then commanding Camp Robinson, sent soldiers and Indian scouts to arrest Crazy Horse with the intent of imprisoning him. This attempt was also unsuccessful, but Crazy Horse surprised everyone by surrendering at Camp Robinson of his own accord. Red Cloud was one of twenty Sioux who left Camp Robinson for Washington with Lieutenant Clark three weeks after the killing of Crazy Horse, evidencing a better relationship with military officials. It was the chief's initial meeting with newly elected President Rutherford B. Hayes.[33]

As a part of a get-tough policy following the Great Sioux War, the government decided to move the Red Cloud and Spotted Tail agencies to the Missouri River. A three-man commission selected a site for Red Cloud's people at the junction of Yellow Medicine Creek (now called Medicine Creek) and the Missouri River near present-day Lower Brule, South Dakota. Workmen quickly erected buildings for the new agency. Ostensibly, the move was intended to reduce the government's shipping costs, as the Indians' rations and annuity goods could be delivered by steamboat. Meeting with President Hayes in Washington in late September 1877, Red Cloud and Spotted Tail secured permission to locate off the Missouri near the southwest corner of the Sioux reservation the following year.[34]

Unfortunately for the Lakotas, Hayes insisted that they spend the upcoming winter on the Missouri River. On 25

October, about eight thousand Oglalas and two thousand Northern Cheyennes left the agency to make the long trip to the river with few rations and a little transportation scraped together by the army. About eighty miles from the new agency, virtually all the northern Indians broke away and continued north, while Red Cloud stated that he would go no farther. Over General Sheridan's objections, the acting commissioner of Indian affairs agreed to allow Red Cloud and his band to spend the winter at their chosen campsite on the White River.[35]

In the spring of 1878, Congress authorized the secretary of the interior to appoint a commission to select another site for the new Red Cloud agency. Agreeing to the general area but not the specific site recommended, Red Cloud and his followers took matters into their own hands and moved toward their preferred location on White Clay Creek, a tributary of the White River that crosses the present-day Nebraska-South Dakota border just south of the town of Pine Ridge. The site lay on a vast prairie cut by small rivers and creeks, which were fringed with willows, cottonwoods, and scrubby pines. The Office of Indian Affairs acquiesced once again, and in October 1878 the Pine Ridge Agency officially became Red Cloud's new home.[36] Perhaps the government's refusal to name the agency after Red Cloud reflected a desire to reduce his standing. In any case, the government's struggle to remake Lakota society would continue in earnest at Pine Ridge.

10 Pine Ridge Agency
1878–1886

Senator Dawes maintains that the trouble at Pine Ridge arises from a conflict between the new and the old order of things—between the power of the chiefs and the power of the law.
—Omaha Daily Bee[1]

While Red Cloud and the Oglalas were busy with their move to Pine Ridge in the fall of 1878, a band of about three hundred fifty Northern Cheyennes fled from their reservation in present-day Oklahoma on 7 September and attempted to return to their homeland in Montana. Pursued by the army, the escapees split into two groups under chiefs Little Wolf and Dull Knife as they made their way across Nebraska. Little Wolf's group continued on to Montana, while Dull Knife's group decided to seek refuge with the Oglalas at Red Cloud Agency, not knowing that Red Cloud's people had moved to Pine Ridge. Soldiers caught up with Dull Knife's group in late October 1878 and took the Indians as prisoners to Fort Robinson (formerly Camp Robinson), where they were confined in an abandoned barracks. Fearing that they would be forcibly returned to Oklahoma, Dull Knife and his group of about one hundred twenty-five Cheyennes broke out of their prison barracks to flee into the Nebraska Sand Hills on 9 January 1879. About half were killed or wounded in fighting with the Fort Robinson garrison. A few managed to reach the Oglalas at Pine Ridge, where they were allowed to remain after surrendering to the authorities there.[2]

The Cheyennes' ordeal had a negative impact on Red Cloud's relations with the federal government. Dr. James Irwin had resigned as agent at Pine Ridge on 1 January 1879.

His successor was Dr. Valentine T. McGillycuddy, who would become the chief's most effective reservation adversary—and who suspected Red Cloud of providing arms and ammunition to Dull Knife's Cheyennes.[3] McGillycuddy's suspicions only added to his initial distrust of Red Cloud, whom the agent perceived as an unreconstructed traditionalist. On 1 May 1879, less than two months after the new agent's arrival at Pine Ridge, Red Cloud sent a letter to "Our Great Father" elaborating his grievances against McGillycuddy, calling for the return of Dr. Irwin as agent, noting that the government still owed the Lakotas compensation for the Black Hills, and demanding the prosecution of white thieves who were stealing his ponies.[4]

Valentine McGillycuddy was thirty years old when he arrived at Pine Ridge on 10 March 1879. He had seen more of the West than many white men twice his age. McGillycuddy had served with Professor Walter P. Jenney's Black Hills geological expedition in 1875 and with Brigadier General George R. Crook's troops in the fall of 1876, and he had administered medical aid to the dying Crazy Horse at Camp Robinson in 1877. While in Washington in January 1879, McGillycuddy met with Commissioner of Indian Affairs Ezra A. Hayt and Secretary of the Interior Carl Schurz to complain about the poor treatment of the Lakotas. Respecting his passion and ideas, these officials offered him a job as Indian agent at Pine Ridge.[5]

Bright, hot-tempered, quick to act, and stubborn, McGillycuddy made it his mission to start his charges on the white man's path through education, Christianization, and agriculture. Determined to assimilate the Lakotas into "civilized" society as soon as possible, he lacked patience and finesse. To accomplish his program, McGillycuddy immediately attacked the Oglala social and political structure. The agent's first step was to break the power of the chiefs—especially that of Red Cloud. One important source of strength for McGillycuddy in his struggles with Red Cloud was the respect the young doctor had earned among Crazy Horse's admirers and followers for his treatment of the dying chief at Camp Robinson. Thus, Crazy Horse's people often supported the agent in his endeavors.[6]

McGillycuddy met Red Cloud and the other Oglala chiefs for the first time shortly after assuming his position in March 1879. McGillycuddy described the meeting in a letter to Elmo Scott Watson almost forty-three years later, recalling that he told Red Cloud, "The white man has come to stay; and wherever he places his foot the native takes a backseat." Red Cloud replied, "It is not right," and to this McGillycuddy responded, "It is not a matter of right or wrong, but of might and destiny."[7] Thus, the lines were drawn, and the battle raged for as long as both men were on the reservation. Red Cloud would make many trips to Washington to complain about the Pine Ridge agent. McGillycuddy may have been single-minded and his methods sometimes pushed the envelope, but no one could prove that he was anything but conscientious in his attempts to apply his ideas to improve reservation life. The cause of contention was simple: Red Cloud wanted to keep the Lakota culture and way of life in isolation—away from the general white population—while McGillycuddy wanted to make the Oglalas live like white people in keeping with the 1868 Treaty of Fort Laramie.[8]

The contest between chief and agent played out on many fronts, one of which was the Indians' means of livelihood. Red Cloud scorned suggestions that Indian men begin farming and ranching and gave Agent McGillycuddy his straightforward view of the matter: "Father, the Great Spirit did not make us to work. He made us to hunt and fish. . . . The white man can work if he wants to, but the Great Spirit did not make us to work. The white man owes us a living for the lands he has taken from us."[9] McGillycuddy, on the other hand, sought to instill his version of self-sufficiency in every Indian at Pine Ridge by promoting farm and ranch culture on reservation lands. Government officials also saw this policy as a means of reducing rations over time.[10]

Another area of contention was land ownership. McGillycuddy planned to settle Indian families on individual homesteads throughout the reservation. The agent saw this policy as a means of diminishing the power of the chiefs, who had the say in a communal society. McGillycuddy described the idea in his 1880 annual report to the commissioner of Indian affairs: "In inducing them to scatter out in this way, I have naturally incurred the ill will of some of the chiefs, as they—the chiefs—are fully alive to the fact that as soon as these Indians become house-owners and land-owners, their glory as petty potentates will have departed. So I have necessarily met much opposition, notably from Red Cloud, who, with the neighboring chief Spotted Tail, form about as egregious a pair of old frauds in the way of aids to their people in civilization as it has ever been my fortune or misfortune to encounter."[11]

McGillycuddy's plan was but one of several concerning the future use of Indian lands, as Dakota Territory's white residents lobbied long and hard to open up more areas for settlement. One key ally of the settlers' interest was former territorial governor Newton Edmunds, who led a three-man commission authorized by Congress to meet with Sioux leaders for that purpose in 1882. The commission's plan was to give the Indians homesteads on what remained of the reservation after its division into five separate units. Edmunds and his colleagues met strong resistance from Red Cloud when they visited the Pine Ridge Agency in October 1882.[12] The chief delivered this message: "We will not sell our land. Why ask us why? I will tell you. . . . We signed away the Black Hills; they were full of gold. Have we ever received an equivalent for them? Look at my people. We have no money; our pockets are empty, and we are poor. This is our land by mutual consent of the Indians and of the whites. We shall never willingly yield it to the whites."[13]

While the Edmunds Commission recommended that the southern portion of the reservation be reduced by over seventeen thousand square miles and Congress considered an agreement to this end, the 1882 proposal did not come to fruition. The Indians' friends in Congress, notably Massachusetts Senator Henry L. Dawes, blocked Senate approval of the document and successfully pressed for a select committee to investigate the matter. Dr. Thomas A. Bland and his magazine, the *Council Fire and Arbitrator*, also were influential in discrediting the Edmunds Commission's report. Although Red Cloud got his way in 1882, the ongoing battle over Oglala lands would continue beyond McGillycuddy's tenure as agent.[14]

McGillycuddy, however, continued to implement policies designed to undermine Red Cloud's authority. One way a Lakota chief traditionally kept his power was by providing for the people's physical needs. Initially, the government's system of issuing rations and annuity goods to reservation Indians reinforced this Lakota tradition, as chiefs personally distributed food and supplies to their followers. McGillycuddy tried to change this pattern, first by having the agent replace the chiefs as the distributor of supplies and, later, by recognizing the leaders of ever-smaller groups as chiefs, thus creating as many as sixty-three annuity dispensers. Red Cloud quickly understood the implications of this last practice and waged a political war over it. He also knew that some of the goods were not finding their way into the tipis and bellies of the Indian people. Much to the agent's chagrin, McGillycuddy discovered in 1882 that the law required that annuities be turned over in their original parcels to recognized chiefs for distribution.[15]

McGillycuddy also used education as a means of promoting white values, establishing day and boarding schools on the reservation and encouraging Oglala parents to send their children to the Carlisle Indian Industrial School in Carlisle, Pennsylvania. Red Cloud opposed these schools as a threat to Lakota culture, but McGillycuddy declared that the chief did not want the young people "to know and respect the strength of the government."[16] Seventy children in Red Cloud's band did not attend the agency day school despite the fact that McGillycuddy sometimes withheld food rations from their families. Red Cloud, not satisfied with opposing the schools on his own reservation, visited Standing Rock and Cheyenne River agencies, where he made his views known. Standing Rock Agent James McLaughlin promptly ordered him home. A few Oglala and Brule children did attend Carlisle, but when Red Cloud and Spotted Tail visited the facility in July 1880, they publicly expressed dismay with the school's military-style regimentation and discipline.[17]

As a means of keeping Oglala children closer to home, Red Cloud supported the Jesuits' 1887 establishment of the Holy Rosary Mission (later Red Cloud Indian School), located four and one-half miles north of Pine Ridge Agency. Jesuit priests, Sisters of Saint Francis, and lay volunteers composed the school's staff. The institution was the culmination of both Red Cloud's numerous requests for Catholic missionaries and the withdrawal of federal restrictions on Christian evangelizing on the reservation. Red Cloud had asked for a Catholic school as early as 1877 and again in 1879. In the latter case, Agent McGillycuddy had been able to keep Catholic missionaries off the reservation, citing official policy that allowed only one designated Christian denomination to do missionary work on each reservation. The Lakotas remembered Jesuit Father Pierre-Jean De Smet, who had ministered to many Plains tribes from the 1830s to the 1860s, and repeatedly requested that Catholics come to their agencies in lieu of Episcopalians, who had been the government's favored denomination at Pine Ridge.[18]

Another means of breaking tribal cohesiveness was to forbid native ceremonies and religious practices. Agent McGillycuddy especially wanted to suppress the Sun Dance, which his 1884 annual report called "a barbarous and demoralizing ceremony, antagonistic to civilization and progress."[19] Red Cloud was a leading participant in an 1882 Sun Dance at Pine Ridge, which the Oglalas held despite their agent's disapproval. McGillycuddy undoubtedly supported the Code of Indian Offenses issued by Secretary of the Interior Henry M. Teller on 10 April 1883, which prohibited the Sun Dance and other religious ceremonies, many practices of traditional medicine men, the holding of potlatches (giveaways), the practice of polygamy, and the obtaining or selling of liquor on the reservations. The new regulations also created on each reservation a Court of Indian Offenses, whose decisions were subject to review by the agent. These courts had power to impose penalties, including the withholding of rations, incarceration, or removal from the reservation. The code, which outlawed several key elements of Sioux culture, was a terrific blow to the Lakota people and to Red Cloud's prestige. Nevertheless, some bands continued to conduct important ceremonies in secret.[20]

Before the Oglalas moved to Pine Ridge, enforcement of the community's norms had been the province of Red Cloud's devoted young warriors. On 27 May 1878, however, Congress provided funds to establish Indian police units on western reservations. When McGillycuddy took over as agent in March 1879, the Pine Ridge police force consisted of four men. While Red Cloud effectively resisted the recruitment of Indian police officers at the beginning of the young doctor's tenure, thieves stole three thousand head of Indian horses that summer, leading to the recruitment of forty-six additional Oglala policemen in August and September. The respected shirt-wearer George Sword agreed to serve as the force's captain.[21] McGillycuddy understood well that the creation of a strong police force undermined Red Cloud's power. In an 1881 letter to a colleague, he wrote: "The Indians generally recognize the police authority, for from time immemorial there has existed among the Sioux and other tribes native soldier organizations systematically governed by laws and regulations. Some of the strongest opposition encountered in endeavoring to organize the police force . . . was from these native organizations, for they at once recognized something in it antagonistic to their ancient customs, namely, a force at the command of the white man opposed to their own."[22] Further reduction of Indian policing power came in 1885 when Congress passed the Major Crimes Act, which assigned jurisdiction on Indian reservations to federal courts in the matter of seven crimes: murder, manslaughter, rape, assault with intent to kill, arson, burglary, and larceny.[23]

The disagreements between the agent and the chief came to a head on 13 August 1882, when Red Cloud and fifty-two other leaders and headmen signed a letter to Secretary of the Interior Henry Teller demanding that McGillycuddy be replaced within sixty days because of "gross misconduct" in office. If no action was taken, the letter suggested, the Indians would "politely escort him out of our country."[24] At that point there were roughly eighty-three hundred Indians at Pine Ridge, including four hundred Northern Cheyennes and six hundred former "hostiles" from Sitting Bull's band, and the small agency police force could not withstand a major attack. In response to the letter, McGillycuddy called a general council on 18 August, meeting with Indians camped near the agency who were known to support him in his tug-of-war with Red Cloud. McGillycuddy told the assembled group that because of the action the disgruntled chief had taken in threatening to remove him, Red Cloud was defying the government. Consequently, the agent stated that he had brought them together to discuss the need for troops to quell the insubordination.[25] The response was what McGillycuddy had expected, and many attendees signed the following letter to the commissioner of Indian affairs:

> Pine Ridge Agency, Dakota
> August 18, 1882
> We the undersigned chiefs, headmen, and Indians
> of Pine Ridge Agency Dakota desire to inform the
> Great Father that we do not require the presence of
> troops here. We blame Red Cloud and other Indians
> who signed the letter of threats recently forwarded
> to you. We agree to settle the trouble with the aid of
> the police and aid the Great Father to prevent and
> settle all trouble in the future.
> (Signed) Little Wound, Young Man Afraid of His Horses,
> and twenty others.[26]

The next day, McGillycuddy received an order from the commissioner authorizing Red Cloud's arrest at the agent's discretion. Several attempts to get the chief to come in peaceably were met with evasion and promises of future talks. Finally, McGillycuddy sent Red Cloud a message that troops would come for him if he did not appear. Soon after, the chief arrived with seven hundred of his personal band. Agent McGillycuddy read the arrest warrant and told the Indian police and the "friendly" chiefs and Indians that he would hold them responsible for Red Cloud's future conduct.[27] McGillycuddy did not actually arrest the chief, however. Red Cloud quickly backed down from the confrontation, but as the *Omaha Daily Bee* stated in 1884, "It should . . . be borne in mind, that as long as Red Cloud lives, there is liable to be more or less trouble among the Pine Ridge Indians, because the deposed chieftain will never neglect an opportunity to make it interesting for McGillycuddy."[28]

For the agent, the August 1882 confrontation with Red Cloud was not a complete triumph, for the controversy resulted in an investigation. The Indian Office dispatched Inspector William J. Pollock, who had no particular sympathy for Agent McGillycuddy. Pollock noted in his first meeting with Red Cloud, "Tears started to the old man's eyes and he put his arms around me, saying that his face had been pressed to the earth for three years."[29] Pollock reported evidence of corrupt practices among supply contractors and sent a horseshoe far too large for an Indian horse to the acting secretary of the interior with this note: "Preserve this specimen brick of our Indian management till I see you."[30] According to the *Salt Lake Daily Herald*'s account of Pollock's find, the horseshoe exemplified "the very loose way of doing business on the part of Indian agents who should not accept such goods."[31]

McGillycuddy apparently complained about Pollock's investigation to the *Omaha Daily Bee*, whose edition of 25 September 1882 accused the inspector of "carrying on his investigation with a high hand" and further declared that "the testimony of every squaw man and half-breed is eagerly accepted."[32] When Pollock finished his field investigation, he peremptorily suspended McGillycuddy from office without the approval of the Interior Department.[33] Secretary of the Interior Henry Teller, however, suspended Pollock and the investigation because, according to the *St. Paul* (Minn.) *Daily Globe*, the inspector "had assumed the powers of secretary of the interior, taking the liberty to willfully disobey Secretary Teller's orders, and substituting his own ideas of dealing with McGillycuddy."[34] Agent McGillycuddy was cleared of the charges made by Red Cloud and Inspector Pollock. In retrospect, the action appears to have been an example of the Indian Office protecting its own.

In December 1882, an uncowed Red Cloud went to Washington to plead his case against McGillycuddy once more. He declared: "I want a new agent; McGillycuddy is a bad man. He is quarrelsome and calls us bad names. He says we are old women. He steals our supplies. You do not make him vouchers for our supplies. You take his word and he steals from us."[35] Red Cloud also used the trip to demand compensation for horses General Crook had taken from the Oglalas in 1876. Commissioner of Indian Affairs Hiram Price reportedly remarked that the seven thousand horses the army had taken from the Sioux had been sold for a total of $19,400, and the War Department could not satisfactorily account for the proceeds. Not to be denied, Red Cloud testified before the House Committee on Appropriations, as well. He eventually received his compensation but not until 1889.[36]

According to the *Omaha Daily Bee*, Agent McGillycuddy's many antagonists included reformers the editor described as "theoretical philanthropists whose whole idea and knowledge of the noble Red man is originally obtained from reading [James Fenimore] Cooper's novels and interviewing Indian tobacco signs," as well as shady traders, contractors, and horse thieves.[37] Former Pine Ridge trader H. C. Dear's accusation that McGillycuddy had offered one hundred dollars to any Indian who would kill Dear because he had been telling the truth about the agent's practices was published in the *Salt Lake Daily Herald*. A reformer who definitely proved to be a significant thorn in the agent's side was Thomas A. Bland of the National Indian Defense Association. Bland and his associates backed Red Cloud's resistance to land cessions. Bland's organization was one of the few non-Indian groups to oppose breaking reservations up into individual parcels, believing that the Indians should decide the land issue by democratic vote.[38]

In 1884, Red Cloud asked Bland to visit Pine Ridge in order to get his opinion on the legislation that Senator Dawes had introduced to divide and reduce the Sioux reservation. When Bland arrived in late June, Agent McGillycuddy had six Indian police officers escort him off the reservation. Bland published his version of this event in the *Council Fire*, where he accused McGillycuddy of falsifying agency accounts for his own financial benefit. Inspector Henry Ward investigated Bland's charges and concluded that McGilly-cuddy had done no wrong. The Board of Indian Commissioners also visited the agency and returned a favorable report about McGillycuddy to the Interior Department.[39]

In the spring of 1885, Red Cloud was in Washington again, complaining that his people were "suffering for food" because they had "not received any coffee, sugar or flour for three months."[40] Red Cloud reported that the people had asked him to tell the Great Father that they wanted a new agent. On 25 April, Red Cloud, accompanied by Thomas Bland and interpreter Todd Randall, met with Commissioner of Indian Affairs John D. C. Atkins. Agent McGilly-cuddy, Young Man Afraid of His Horses, and two Indian police officers were also present. Endeavoring to crush the agent and destroy his influence, Red Cloud presented a written statement in which he charged McGillycuddy with misappropriating funds. McGillycuddy entered a general denial. The chief lobbied for McGillycuddy's dismissal, and President Grover Cleveland eventually agreed to investigate the matter.[41]

Taking advantage of the fact that Congress had directed the speaker of the House to appoint a five-man committee to investigate expenditures for Indian affairs, the Cleveland administration referred the troubles at Pine Ridge to the panel, chaired by Representative William S. Holman of Indiana. Holman's committee visited Pine Ridge in July 1885 and took testimony from both sides. The committee's report of 18 March 1886 did little to resolve the controversy between Red Cloud and Agent McGillycuddy. Indeed, the report mentioned McGillycuddy only in its record of testimony and made no recommendation either to retain or remove him.[42]

As the controversy continued into 1886, McGillycuddy's supporters maintained that the agent's accusers had ulterior motives. Elaine Goodale, who had visited Pine Ridge earlier that year, concisely summarized this viewpoint in a letter that ran in the *New York Evening Post* and elsewhere:

> There are then at this great agency two fully organized parties—on the one side this remarkable man [McGillycuddy], all coolness, nerve and executive force, with a backing of fifty Indian police, well armed and disciplined, under Captain [George] Sword, an almost equally remarkable Indian, and a majority of the chiefs with their bands of followers; on the other the famous old malcontent Red Cloud, obstinately fighting for his declining influence, and surrounded by a little band of dissatisfied, turbulent and non-progressive Indians. The "Red Cloud faction" is a refuge for every Indian with an old enmity toward the government or a personal grudge against the agent; every lazy Indian who doesn't want work; every objector to the schools, who wants to keep up Indian dances, dress and customs; in a word, all the chronic grumblers and "coffee-coolers," and the whole "opposition" element.[43]

Finally, McGillycuddy's superiors in the Indian Office had enough of the bickering, charges and countercharges, threats, and confrontations emanating from Pine Ridge. In May 1886, Grover Cleveland's Democratic administration ordered McGillycuddy, a Republican appointee, to replace his efficient Republican clerk. As the Indian Office expected, McGillycuddy refused to comply and was dismissed. Red Cloud had finally won.[44] The *Omaha Daily Bee* spoke for those who regretted McGillycuddy's departure: "He has quelled every sign of disorder among the Ogallala Sioux and Cheyennes, fought off the gang of dishonest traders and contractors and maintained his position against the host of enemies which his honest and efficient management of agency affairs brought down upon him. Under his vigorous if arbitrary rule, the Sioux have been kept at peace, factionalism has been made harmless, schools have been built and filled with children, and many of the Indians have become largely self-supporting."[45] There was no doubt that Valentine T. McGillycuddy had left his mark and no chance that Red Cloud would soon forget him.

11

Pine Ridge Agency
1886–1893

Just as sure as you once lived on the buffalo,
you must in the future live as the white man lives.
—Major General George R. Crook[1]

After Valentine T. McGillycuddy left Pine
Ridge, Captain James M. Bell of the Seventh Cavalry temporarily replaced him as agent. During his four–month
stay, Bell got along well with Red Cloud. The chief sang
his praises in a letter to Dr. Thomas A. Bland: "Since the
arrival of Capt. Bell here none of my people have had cause
to complain, as he has adopted the just and manly course of
treating all Indians alike, without regard to former cliques
and clans. . . . [Bell] has restored to my people all the ration tickets that had unjustly been taken from them by Mc-
Gillycuddy." Red Cloud claimed that the former agent had
withheld about nine hundred tickets. Some of the affected
Indians had not drawn rations for as much as three years
and depended on friends for food.[2] Bell, in turn, praised the
Indian police highly, calling them "very efficient" and "entirely trustworthy."[3] Five months after McGillycuddy's dismissal, Hugh D. Gallagher of Greensburg, Indiana, became
the permanent agent at Pine Ridge. Like Captain Bell before
him, Gallagher established a rapport with Red Cloud and
the other chiefs.[4]

The first controversy at Pine Ridge under Agent Gallagher's administration would come via the General Allotment Act, also called the Dawes Act after its congressional
sponsor, Massachusetts Senator Henry L. Dawes. Signed by
President Grover Cleveland on 8 February 1887, the legislation marked the culmination of government efforts to settle Indian families on separate tracts of land and replace
the concept of communal property with that of individual

ownership. The legislation authorized the president to order land surveys on the affected Indian reservations, after which these lands would be divided into parcels for distribution to Indian families and individuals. Heads of families were to receive 160 acres, single adults and orphaned children would get 80 acres, and other children would receive 40 acres each. Under an 1891 amendment to the act, allotments of land suitable only for grazing were doubled in size. The law required the federal government to keep allotted parcels in trust for twenty-five years, at which point the Indian owners received full legal control of their individual allotments and became United States citizens. The Dawes Act further provided that reservation lands not needed for individual allotments were to be purchased from the tribes by the federal government and opened for settlement. Settlers had already used the Homestead Act (1862) and other federal laws to acquire most of the land suitable for agriculture in the West, hence, "surplus" Indian lands became the agrarian settlers' last frontier.[5]

The next step in implementing national Indian policy in Dakota Territory was to reduce the size of the huge Great Sioux Reservation, the effort spurred on by land-hungry interests. On 30 April 1888, Congress passed a Sioux Bill, which proposed dividing the reservation into six smaller units and selling "surplus" lands to settlers for fifty cents an acre, the proceeds to be deposited into an Indian trust fund. Under the 1868 Treaty of Fort Laramie, the government's proposal needed the approval of three-fourths of the adult men living on the reservation to take effect. Red Cloud firmly opposed the proposal, as did most of his fellow Lakotas and Thomas Bland's National Indian Defense Association.[6] At a council of chiefs and prominent men at Pine Ridge Agency in late May, sentiment was against surrendering the land. The *Omaha Daily Bee*'s account of the meeting described Red Cloud as the "grand mogul" among the assembled chiefs.[7] The government brought a Lakota delegation to Washington in October 1888 in an attempt to drum up support for the land cession, but Red Cloud was not among them because of his opposition. Even without Red Cloud, the Lakota delegates made it clear that the 1888 Sioux Bill was unacceptable.[8]

The Indians' arguments achieved some success in that Congress eventually passed a revised Sioux Bill. The new bill, signed into law by the outgoing President Cleveland on 2 March 1889, still authorized dividing Sioux lands into six smaller reservations and selling the "surplus" land to settlers, but it raised the price from fifty cents to $1.25 per acre for land sold in the first three years of the process. As a further concession, Red Cloud, Red Leaf, and their followers were to be paid forty dollars' compensation for each horse the army had taken from them in 1876. President Benjamin Harrison appointed former Ohio governor Charles W. Foster, former Missouri representative William Warner, and newly promoted Major General George R. Crook as commissioners to secure Lakota consent to the new Sioux Act on 19 May 1889.[9]

President Harrison's commissioners arrived at Pine Ridge on 13 June, having already collected Lakota signatures approving the Sioux Act at Rosebud Agency. Red Cloud again led the Pine Ridge opposition to the land-taking proposal, arguing that his people needed all their present land holdings if they were to adapt successfully to white civilization. The chief's address to the commissioners on 18 June focused on the promises the government had made to the Lakotas in past treaties and agreements, many of which had not been fulfilled. Under the treaty of 1868, he said, the Lakotas were to receive cattle, horses, sheep, and farm implements. Since that time, Red Cloud reminded the commissioners, the Indians had given the Great Father the Black Hills, and, as he understood the agreement, the Lakotas were to be paid for seven generations.[10] "Now, my friends," he declared, "the Great Father has not paid the things promised us, but wants us to give more land before we are paid for what is due. I looked around to see if you had any boxes full of money to pay us, but I see none. I presume you are to pay us in sugar talk as you have done before."[11]

Despite Red Cloud's outspoken opposition to the land cession, the commissioners obtained signatures from three-fourths of the adult male residents of the Great Sioux Reservation by late July 1889. The official tally was 4,482 signers out of 5,678 eligible men across the reservation. General Crook, in particular, had effectively argued that the government's terms were the best the Lakotas were likely to get. Even so, only 684 out of 1,306 adult males at Pine Ridge Agency signed the agreement. In the end, the government would take another nine million acres and divide the former Great Sioux Reservation into six smaller units: the Cheyenne River, Lower Brule, Rosebud, Crow Creek, Standing Rock, and Pine Ridge reservations.[12] After the commissioners left Pine Ridge, Red Cloud told a reporter that even though the 1889 Sioux Bill was "the best the Great Father ever offered us," it was still "not good enough" for him because it did not "give us enough for our land." The chief said further that he expected the land to increase in value over time, so he wanted to wait before selling any of it.[13]

On 2 March 1889, at the same time it passed the Sioux Act, Congress dealt the Indians another severe blow, cutting the 1890 fiscal-year appropriation "for subsistence and civilization of the Sioux" by 10 percent from the level of the previous two years.[14] The cut, amounting to one hundred thousand dollars, caused a steep reduction in the annual beef ration at Pine Ridge and Rosebud. Indian Office records show that the Pine Ridge beef ration was reduced from five million to four million pounds, while Rosebud's beef supply fell from over eight million to six million pounds. The Indian Office informed the agents at Pine Ridge and Rosebud of the ration cut in June 1889—while General Crook and his fellow commissioners were still collecting signatures on the

Sioux Act land-cession agreement. The Indians were apparently not told of the ration cut until after the commissioners left. The reduced rations caused great hardship. In December 1890, Red Cloud informed Bland that over two hundred people had starved to death since the fall of 1889 and that many others were sick due to insufficient food. Two seasons of dry weather had hampered the Indians' efforts to raise crops, and the meager rations had forced them to kill what cattle they had.[15]

Later amendments to the Dawes Act enabled whites to acquire or to control land that had originally been allotted to Indians. The first such change occurred on 28 February 1891, when Congress authorized the secretary of the interior to determine whether an individual Indian was capable of occupying or improving his land. If the secretary determined that the Indian owner could not effectively use the allotment, the land could be leased to another party on terms negotiated by the Interior Department.[16] Six years later, a Pine Ridge Indian council sent Red Cloud, American Horse, Clarence Three Stars, and Patrick High Star to Washington to present Lakota concerns to Congress. A newspaper account described Red Cloud, then about seventy-six years of age, as "a pitiful spectacle" and "almost blind and very feeble" as he appeared before the Senate Committee on Indian Affairs. Nonetheless, the aged chief spoke resolutely against the allotment policy: "We do not want our lands allotted to us in severalty [under the Dawes Act]. We are willing to become farmers, as our white brothers demand, but the lands are bad and fit only for grazing and raising cattle and horses. We have tried to plant, but the sun in July and August scorches everything. There is not enough water for agricultural life. There are only a few places where even the cattle can get a living in summertime."[17] His words fell on deaf ears.

With legislation passed in May 1900, Congress gave the secretary of the interior broader discretion to determine whether allottees were capable of using their land in the manner the government desired. Again, the Interior Department was empowered to arrange leases for lands held by Indians whom the department judged incapable of using them properly.[18] In 1906, Congress passed the Burke Act, which authorized the secretary of the interior to issue a patent in fee simple to those Indians classified as "competent and capable." This authorization meant that allottees deemed competent would have their land taken out of trust status, thus becoming subject to taxation. These Indians were now free to sell their land to anyone they wished. In many cases, allottees quickly sold their lands to whites.[19]

The near-extermination of the buffalo herds by 1882, the prohibition of Indian religious practices in 1883, the loss of judicial autonomy under the 1885 Major Crimes Act, the loss of lands under the 1887 Dawes Act and the 1889 Sioux Act, the reduction in rations in 1889–1890, epidemics of measles, influenza, and whooping cough, and general repression finally had their effect. Desperation led to yearning, and deep depression birthed religious revival. Out of this crisis grew the so-called Ghost Dance movement.[20] Anthropologist Jack Goody has pointed out that in times of stress, when one civilization is hard pressed by another, a tendency exists "to seek comfort in ritual designed to bend time backward" or to accelerate it forward, to seek either an earlier Garden of Eden or the coming of the millennium.[21] The late-nineteenth-century Sioux looked both ways in their search for means to drastically change their present. Red Cloud succinctly expressed the temper of the moment: "There was no hope on earth, and God seemed to have forgotten us. Some said they saw the Son of God; others did not see Him. If He had come, He would do some great things as He had done before."[22]

In the late 1880s, the most well-known medicine man preaching religious revival to his fellow Indians was a Paiute named Wovoka, known to whites as Jack Wilson. Wovoka's teachings promised a future in which *Wakan Tanka* (variously translated as the Great Mystery, the Great Spirit, or God) would free Indian people from Euro-American domination if the Indians adhered to proper morals and performed the requisite rituals, including what came to be known as the Ghost Dance. Wovoka promised his followers that they would once again see their deceased relatives, the buffalo would return from the verge of extinction, and *Wakan Tanka* would remove whites from the world through some sort of cataclysmic event. Wovoka told his followers not to fight the whites; instead, they were to leave Euro-Americans' fate to *Wakan Tanka*. Even so, many Euro-Americans saw the movement as a threat.[23]

In March 1890, Short Bull, Kicking Bear, and nine others who had traveled to Nevada to meet Wovoka returned to the Pine Ridge, Cheyenne River, and Rosebud reservations to spread the message they had received. These emissaries encountered resistance from their reservations' agents, who sought to prevent the spread of the Ghost Dance and, by the fall, ordered dancers to stop. Given that the government had already attempted to suppress the Sun Dance and other Lakota religious practices, the agents' hostility to the Ghost Dance was hardly surprising. The number of Sioux people who took up the Ghost Dance was a surprise to the agents, however. On 22 August, Agent Hugh Gallagher ordered his Indian policemen to stop a Ghost Dance on the Pine Ridge reservation. About two thousand people were in attendance as either participants or spectators. The dancers ignored Gallagher's order to stop, and the police could do little because they had not been authorized to use force. By one scholar's estimate, between one-fourth and one-third of residents on the Rosebud, Pine Ridge, Standing Rock, and Cheyenne River reservations were involved with the Ghost Dance movement at its peak in the fall of 1890.[24]

On 1 October, at a time when the appeal of Wovoka's teachings among the Oglalas and Brules was at its peak,

Daniel F. Royer replaced Hugh Gallagher as agent at Pine Ridge. Royer, a new and inexperienced agent whose political connections were his sole qualification for office, proved to be the catalyst for tragedy. Four weeks after his arrival, frightened and completely unable to cope with Ghost Dancers who defied his orders to stop, he dispatched a frantic plea for military protection. On 20 November, the first contingents of troops from Omaha and Forts Robinson and Niobrara in Nebraska arrived at the Pine Ridge and Rosebud reservations. By the end of the month, thousands more from surrounding states had arrived on the scene. Major General Nelson A. Miles, commanding the operation, concentrated his largest force at Pine Ridge. About three thousand troops were sent there, including the entire Seventh Cavalry Regiment under Colonel James W. Forsyth.[25]

At the appearance of the troops, Kicking Bear, Short Bull, and three thousand Ghost Dancers fled to the badlands in the northwest corner of the Pine Ridge reservation, about fifty miles northwest of the agent's headquarters. In the meantime, Minneconjous under Big Foot left the Cheyenne River Indian Reservation and started for Pine Ridge to join their Oglala relatives. On 28 December, the Seventh Cavalry intercepted Big Foot's party near Wounded Knee Creek. Some of the Indians resisted being disarmed the next morning, and in the resulting conflict, the soldiers killed 175 or more men, women, and children. The army's casualties were 25 men killed and 39 wounded.[26]

The *Pittsburg Dispatch*'s account of events at Wounded Knee gave an explanation for the killing of Indian families: "As for the squaws, they were not killed with particular intent, notwithstanding that they had been running around with scalping knives trying to stab the soldiers. They were killed principally because they became so mixed with squads of bucks that made dashes to gain the ravine, and were mowed down by the [artillery] battery."[27] Black Elk, who survived the tragedy, saw it another way: "Many were shot down right there. The women and children ran into the gulch and up west, dropping all the time, for the soldiers shot them as they ran. . . . The snow drifted deep in the crooked gulch, and it was one long grave of butchered women and children and babies, who had never done any harm and were only trying to run away."[28]

Red Cloud's role in the events that followed is unclear. The *Pittsburg Dispatch* reported that Two Strike, Little Wound, Short Bull, and other chiefs, with "hundreds of warriors," left Pine Ridge Agency after hearing of the events at Wounded Knee, "compelling Red Cloud to accompany them under threat of death."[29] The chief's son, Jack Red Cloud, was reportedly among the Ghost Dancers.[30] In a letter to Bland dated 12 January 1891, Red Cloud stated: "The Brules forced me to go with them. I being in danger of my life between two fires, I had to go with them and follow my family. Some would shoot their guns around me and make me go faster."[31] On 3 January 1891, the *Omaha Daily Bee* told its readers, "Old Red Cloud tried to slip away from [his captors] and return to the agency," but when they discovered his escape attempt, "they shot all of his ponies, numbering about fifteen, placed a guard over the old chief and proceeded to move several miles further from [the agency]."[32] Years later, Red Cloud's descendant Charging Girl related that the chief was in a difficult position during the Wounded Knee trouble: "After he was shot at by soldiers, he fled with his family to [Joe] Merrivale's place," and later "Indians shot at him, too, because he had not openly sided with them."[33] Red Cloud reached the army's lines on 9 January, having walked sixteen miles.[34]

One interpretation of Red Cloud's role in the aftermath of Wounded Knee that fits the facts is that while he was sympathetic to the Ghost Dance movement, he did not favor violence from either the dancers or the army. On 3 December, he had sent his son Jack to escort a Catholic priest who tried to mediate the situation, and he continued to relay information to government officials while recommending that they take a patient approach with the Ghost Dancers.[35] This interpretation is borne out by Lieutenant Charles W. Taylor's account of meeting Red Cloud just after the chief's escape from his captors:

> Through the scouts daily touch was kept with the hostile camp [in the Badlands]. . . . Red Cloud sent word that he was ready to come as soon as he could escape from camp unnoticed. . . . It was a miserable night following a wet snowstorm—cold, raw, with a heavy damp mist prevailing. About dawn . . . a stealthily moving figure could be distinguished against the skyline. In [the] course of time the old chief, assisted by scouts, staggered into camp. He was placed on a box in my tent close to a Sibley stove in which a fire burned. A squaw and a young girl had accompanied him, as his eyesight was failing him and he experienced difficulty in walking alone. Certainly he was a wreck. All in a tremble, cold, wet, exhausted, hardly able to articulate. . . . I gave him a drink of good whiskey. . . . [The liquor] was magical in its effects, for in a few minutes a wonderful change took place. . . . Light came into his eyes, and quickly he began to talk, his voice increasing in strength. . . . Red Cloud then proceeded to describe the situation among the hostiles in full, giving as his opinion that they would soon give up and come into the agency. This proved to be correct.[36]

The Wounded Knee tragedy largely marked the end of Red Cloud's career as a political force among the Oglala people. As his health failed in later years, he was less able to participate in reservation affairs. Although Red Cloud was part of one last delegation to Washington in 1897, his words were little heeded. Indeed, newspaper coverage of his statement to a Senate committee emphasized his frailty as much as his message.[37] Increasingly, the old chief withdrew into retirement at his home near Pine Ridge Agency, where he spent his final years.

12 Last Years

Fame yet will shout aloud
His praises, who points to righteous peace
And waits, bold brave Red Cloud.
—Lydia H. Tilton[1]

Across from the Pine Ridge boarding school stood a two-story frame house girdled by some desolate-looking tipis, several sweat lodges, and a few log buildings. The camp was Red Cloud's; the house, the only two-story building at the agency, was his residence.[2] The house contained chairs, dishes, and tables where Red Cloud kept all his letters and newspapers "for his children to read."[3] In summer, he pitched his tipi nearby and built a sun shade from upright willow poles covered with leafy branches. In this house, Red Cloud spent his remaining years, infirm, blind after 1897, and largely forgotten. His twenty-one visits to Washington behind him, he now lived quietly on a reservation that no longer violently rejected the will of the dominant society that surrounded it. While many Oglalas resisted the white man's path, they had little recourse against it other than to keep to the old ways in secret.[4]

One of Red Cloud's last outings was a June 1894 hunting trip on the plains near Casper, Wyoming, where his warriors had defeated the army in the Battles of Platte Bridge and Red Buttes in 1865. Accused of killing game out of season by the local sheriff, Red Cloud, his son Jack, and Dreaming Bear were arrested and fined twenty dollars each, which they refused to pay. Red Cloud eventually surrendered his team, wagon, and harness in lieu of the fine and court costs.[5] The victor of Platte Bridge and Red Buttes now experienced a painful humiliation in the same place. As the whites saw it, the centaur of the plains had become an impecunious poacher.

Red Cloud remained an icon of the earlier age of frontier expansion and Indian resistance, however. After his final journey to Washington in 1897, the old chief made a side trip to New York's Madison Square Garden, where he and American Horse witnessed a performance of Buffalo Bill's Wild West from a private box. Afterward, Red Cloud posed for photographs with the show's impresario, William F. ("Buffalo Bill") Cody, and other members of the cast.[6] At a 1902 reenactment of the Battle of the Little Bighorn at Pine Ridge (starring Charles Eastman as Crazy Horse), Red Cloud's followers carried the old chief three miles to the site, where all paid him deference for his years of counsel and leadership. According to a newspaper report of the event, Red Cloud was "totally blind and very feeble." During the reenactment, he "lay under a leafy shelter, his muttering lips showing plainly that he was recalling the events of his own life."[7]

On 4 July 1903, Red Cloud abdicated his position as chief of the Bad Face band to his son Jack and gave a speech that harkened back to the killing of Bull Bear:

> My sun is set. My day is done. Darkness is stealing over me. Before I lie down to rise no more, I will speak to my people. . . . The Great Spirit made us, the Indians, and gave us this land we live in. . . . Then the white man came to our hunting grounds, a stranger. . . . With his trinkets he bought the girl I loved. He brought the *maza wakan*, the mysterious iron that shoots. He brought the *mini wakan*, the mysterious water that makes men foolish. I said, "The white man is not a friend, let us kill him." Our chief, Bull Bear, made me feel ashamed before our people. For the white man he had a heart like a woman.[8]

On 21 September 1903, Red Cloud made his last appearance at a tribal council. The meeting focused on Lakota efforts to gain compensation for the Black Hills, a matter the people had frequently discussed amongst themselves since the 1890s. Now the Lakotas were making their case to public officials, in this instance United States Congressman Eben W. Martin of South Dakota, who had agreed to meet with them at the urging of the commissioner of Indian affairs. Red Cloud, who had the honor of opening the discussion, noted that the Lakotas had been offered six million dollars for the Black Hills in 1875, a sum the old chief likened to "just a little spit out of my mouth."[9] While the approach to Congressman Martin yielded few results, the meeting can be seen as an early step in the decades-long process through which the Lakotas eventually secured legal representation and won a court judgment declaring that they were entitled to compensation for the Black Hills. It would be up to future Lakota leaders to see the process through.[10]

In early 1905, Red Cloud surprised everyone by accepting his land allotment under the Dawes Act. The *St. Louis Republic* recounted the story with its sad overtones:

Eighty-six years old, almost blind, scarce able to hear, broken in health and spirit, well-nigh penniless, Red Cloud, the celebrated chief of the Sioux, once lord of all Kansas, Nebraska and Dakota, and parts of Iowa, Wyoming, and Montana, has at last, after years of bitter and relentless opposition, passed the pipe of peace to Uncle Sam and buried the hatchet, which in his earlier years he drew so often, and, in his age, has brandished so menacingly in war upon what he conceived to be the most cruel and ungenerous foe of the red man. The other day the aged chief tottered into the office of Indian Agent [John R.] Brennan and did what for years he had persistently and stubbornly refused to do namely— accepted his allotment of land. He not only accepted it, but he requested very humbly that it be given to him. At the same time he took occasion to say that he had made up his mind that the allotment policy of the Government, to which for years he had strenuously objected, and which for years he had denounced in unmeasured terms, was a wise policy, and that he had been foolish in opposing it.[11]

As the years passed, Red Cloud's health continued to decline, and his best memories were of an earlier time. Anthropologist Warren K. Moorehead's 1914 account of a conversation with Red Cloud at Pine Ridge may provide a window into the chief's mind in his later years. By his own admission, Moorehead did not quote Red Cloud's precise words because some twenty years had passed since the interview took place. Nonetheless, the anthropologist's approximate record of the old chief's statement rings true:

> Think of it! I, who used to own rich soil in a well-watered country so extensive that I could not ride through it in a week on my fastest pony, am put down here! Why, I have to go five miles for wood for my fire. Washington took our lands and promised to feed and support us. Now I, who used to control 5000 warriors, must tell Washington when I am hungry. I must beg for that which I own. If I beg hard, they put me in the guardhouse. We have trouble. Our girls are getting bad. Coughing sickness every winter [likely tuberculosis] carries away our best people. My heart is heavy, I am old, I cannot do much more. Young man, I wish there was someone to help my poor people when I am gone.[12]

Death came to Red Cloud at Pine Ridge on 10 December 1909. He was buried the following day in the cemetery of the Holy Rosary Mission. Because Red Cloud and his wife had been baptized as Catholics, the chief's funeral was conducted according to the rites of his adopted church. James H. Cook, a white friend, supplied his burial suit. In the end, "civilization" had enveloped him.[13]

In death, Red Cloud received considerable praise in newspaper obituaries and from other writers. The *Valentine* (Nebr.) *Republican* declared: "Red Cloud was the last of the great Indian chieftains. He belongs in a class with Black Hawk, Tecumseh, Logan and Red Jacket. He was never civilized. He fought his battles for his principles. He believed that he had been greatly wronged, and many of his old foes of the wars of 1865–68 will ascribe to that belief."[14] The *New York Times* labeled him "the boldest and fiercest of the Sioux leaders" and a diplomat of rare ability.[15] Even Valentine T. McGillycuddy, the chief's old adversary at Pine Ridge, expressed posthumous respect for Red Cloud, writing ten years after his death: "One could not but admire him for his loyalty to his people and his hatred of the whites. It was simply human nature."[16]

In considering Red Cloud's legacy, it is important to remember that he earned a position of influence among the Lakotas though his exploits on the war trail as a young man. Red Cloud's warrior reputation had been earned in battles against traditional enemies of his people such as the Crow and Pawnee Indians.[17] In the same letter in which he praised Red Cloud's steadfast loyalty to his people, McGillycuddy described the warrior's rise to power: "He was of plebian origin, his parents ordinary Indians, while the Man-Afraid family had been chiefs for generations, and were from father to son hereditary head chiefs of the Ogallalas. Red Cloud starting out as an Indian soldier by fighting ability, won his way up to head soldier, war chief, etc., to head war chief, and after one of the wars assumed 'military dictatorship' and side tracked Man-Afraid."[18] Red Cloud's stature as a military strategist in war against the United States is also beyond dispute. As Warren Moorehead commented in 1908, War Department records then "contain[ed] more frequent mention of Red Cloud than of any other American Indian."[19] At the height of his power in 1867, the *Omaha Weekly Herald* had this to say of him: "The great leader of the Indians is

Red Cloud . . . who is represented as one of the ablest Indian warriors of all time . . . he who conquers Red Cloud will do more than he who conquered Tecumseh, or Black Hawk, or Osceola."[20]

It was also true that after Red Cloud signed the Treaty of Fort Laramie in 1868, he fought no more. At a church reception in Washington in 1889, Red Cloud said: "When I fought the whites I fought with all my might. When I signed a treaty of peace I meant to do right, and I have often risked my life to keep the covenant."[21] He kept his word.

13

Conclusions
Red Cloud's Mind

I further told him that every Indian agent who had been in charge of him and his people— Indian traders and agency employees, as well as white men who had married into his tribe—might all tell conflicting stories about his characteristics, actions, and rating among his people since they had known him, and repeat as facts incidents of his early life told to them, perhaps by forked tongues.
—*James H. Cook*[1]

In looking back at Red Cloud's life, we can say that he was an outstanding warrior, a charismatic leader, a skilled negotiator, and a crafty statesman. He was courageous, generous, creative, farseeing, realistic, and occasionally conciliatory. He was often evasive, obfuscating, and confrontational. He was an expert at procrastinating, and he was sometimes petulant, unreasonable, and incomprehensible. While he led the Oglalas on a new path after the 1868 Treaty of Fort Laramie, Red Cloud remained fully committed to retaining his people's traditions and cultural cohesiveness to the maximum extent possible. In so doing, he resisted the government's efforts to transform his people from equestrian hunters to sedentary agriculturalists almost to the end of his days.

From his own perspective, Red Cloud had ample reason to object to the massive changes the United States imposed on his people during his lifetime. He believed that the Great Spirit had created both Indians and other races, but Indians were created first in the place where they lived, and it belonged to them.[2] Red Cloud's model of proper Indian-white relations harkened back to the fur-trade days. As he remarked to a New York audience in 1870, "I was brought up among the traders, and those who came . . . in the early times treated me well and I had a good time with them."[3] Historian Sylvia Van Kirk has pointed out that in the early fur trade, "white and Indian met on the most equitable footing."[4] Traders did not seek to subdue Indians, take their land, or forcibly change their culture.

As competition for the Indian trade increased, however, whites introduced liquor to the Oglalas and other Plains tribes. Red Cloud grew to despise the presence of alcohol among his people. Obviously his father's death from alcoholism affected his feelings, as did incidents of drunken violence that eroded tribal cohesion, such as the clash between the Bull Bear and Smoke bands in 1841. The spread and use of alcohol remained an important grievance against Euro-Americans throughout Red Cloud's life.

While white men in Oglala country during Red Cloud's youth were interested principally in obtaining beaver skins and buffalo robes, those who came in the mid-1840s and after began to affect the Indians' use of the land. Emigrant wagon trains in the 1840s and 1850s cut a wide swath across the mid-continent, killing and scaring off the Indians' game, depleting the grass, and stripping the trees from the river valleys of the semi-arid Great Plains. Soldiers supplanted fur traders at Fort Laramie in 1849, and government officials tried to define the limits of the Sioux domain with the treaty of 1851. One way for the Lakotas to adapt to this pressure was to move away from it, which they tried to do for much of the 1850s by withdrawing north and west into the Powder River Country and wresting control of that territory from their old rivals, the Crows.

By the 1860s, however, even the Powder River Country was no longer a sanctuary for the Lakotas. Once again, soldiers came to fortify emigrant trails, this time traveled by miners seeking gold in Montana. Having nowhere else to go, the Lakotas chose to fight, resulting in Red Cloud's War of 1866–1868. "We tried to hold our country," Red Cloud declared in an 1878 speech. "We saw white men taking it and killing the buffalo and bringing starvation upon us and our children. This is what made us fight and kill white men."[5]

With the 1868 Treaty of Fort Laramie, the United States government placed binding restrictions on Lakota movements for the first time and began attempts to control the Indians' manner of living. While Red Cloud agreed to the treaty, he later claimed that he had not understood some of its conditions. For example, Red Cloud's 1870 New York speech maintained that he and his fellow Lakota chiefs had not been told what they were really signing two years before. They thought the treaty was simply to remove the forts on the Bozeman Trail and to end hostilities. Only when he had reached Washington, Red Cloud said, did the president explain the treaty's full implications.[6]

While Red Cloud saw many things wrong with the treaty process on his Washington trip, he also came to understand the federal government's ability to enforce its will. Perhaps the most telling circumstance was the great numbers of whites he encountered on the Atlantic seaboard. Red Cloud expressed his wonderment to Secretary of War William W. Belknap. "Our nation is melting away like the snow on the side of the hills where the sun is warm," he said, "while your people are like the blades of grass in the spring when the summer is coming."[7] Red Cloud also was reported to be impressed by the heavy artillery at the Washington Navy Yard.[8]

Experiences from his travels in the East may have gone far to persuade Red Cloud that armed resistance to the government's demands would not be in the Lakota people's interest. Demonstrations of the government's power, however, never convinced the chief that its demands were just. As he remarked in an 1884 letter, "I believe the same Great Spirit made us all, and He doesn't care whether our skin is red, black, or white. What is good for the white man is good for the red, and he has no more right to stand over us in civil life with a gun than he had to crack his whip over the back of a [black] slave."[9] Under intense government pressure to change their way of life, the Lakotas began their transition to the reservation in the 1870s, Red Cloud's objections notwithstanding. The old chief described the process succinctly in a 1903 speech: "The white man came and took our lands from us. They put bounds [on us] and made laws for us. We were not asked what laws would suit us. But the white men made laws to suit themselves and they compel us to obey them."[10]

On the reservation, government officials told the Lakotas that they must become self-sufficient by learning how to farm, which created a number of problems. In perspective, it seems ludicrous to ask a nation of hunters to become plowmen. The Lakotas did not have an agricultural tradition, and their culture glorified the hunt.[11] Lakotas believed, as Red Cloud put it in 1903, that the Great Spirit "gave us the buffalo, the antelope, and the deer for food and clothing."[12] Lakota men saw farming as an unworthy pursuit because it lacked the emotional highs, the adrenalin rush, the immediate satisfaction, and the praise and recognition from fellow tribesmen that hunting could provide. Simply put, all things being equal, farming was boring, demeaning, and irrelevant to Oglala life for most of the nineteenth century.

As the century progressed, all things did not remain equal. Hemmed in on reservations with their land base shrinking, the buffalo disappearing, and dependent on meager government rations, the Oglalas eventually tried to raise food for themselves. By September 1883, the *Washington Post* reported, even Red Cloud had taken to agriculture. The chief owned over one hundred head of cattle and was planning to plow a hundred acres the following year.[13] Lack of agricultural knowledge, however, proved to be a significant obstacle. As Red Cloud told the *Council Fire* in 1884:

We want to farm but don't know how and we have no one to show us. We have small patches dug up and planted the best we know. We don't get seed enough to plant a garden. We want to make hay but don't know how to run a machine. The agent gave us a machine last summer and told us that if we let [whites or mixed-bloods] have the machine he would take it from us and stop our rations. The white men that live with us are good people, and would show us how to work, but the agent won't let them do it. Give us some one to show us how to work and we will do it. We are able and willing but don't know how.[14]

Adding to the Oglalas' difficulties, the Office of Indian Affairs expected them to cultivate essentially barren land in a semi-arid climate. Agent Valentine T. McGillycuddy described the situation at Pine Ridge in his 1881 annual report: "The fact is, that by degrees, the white man has taken from the Sioux pretty much all the land that can be considered arable. . . . White men well trained in farming have tried to till the soil in this vicinity in northern Nebraska and have lost all the money invested, and have not produced enough to pay for the seed."[15] A few years later, the Woman's National Indian Association put the problem succinctly: "[The Indian] has been banished to wild reserves and required to farm where farming would be impossible even to white men."[16] Red Cloud reaffirmed this view in 1897 when he told a Senate committee, "We are willing to become farmers, as our white brothers demand, but the lands are bad and fit only for grazing and raising cattle and horses."[17]

Even if an Oglala farmer were able overcome the obstacles of culture, knowledge, and climate to create an agricultural surplus, the reservation system would not allow him to benefit from it. With few exceptions, an Indian could not sell anything he raised or manufactured on the reservation to anyone other than the traders the government appointed for the purpose—and he often needed the agent's permission to do even that.[18] Red Cloud described the problem clearly in 1884: "The government gives us cows. We take care of them and raise calves, but we can't do anything with them. We can't sell a steer nor eat him, so it is no use to have cattle if we can't sell one or eat one when he gets big."[19]

With all the barriers against self-sufficiency in place on the reservation, it is little wonder that Red Cloud responded at times by telling his people to eschew the system. For example, Joseph Nimmo, a Washington official who visited Pine Ridge, reported in 1885 that Red Cloud believed it was unwise for the Oglalas to work like white people.[20] In August 1903, the old chief argued that the government owed the Oglalas a living in compensation for their lost hunting grounds. Red Cloud suggested that the young men should refuse to work in an effort to convince whites that the Oglala cause was just.[21] In an era when strikes were common in the struggle between capital and labor, Red Cloud proposed one of his own.

Unlike other great Lakotas such as Crazy Horse and Sitting Bull, Red Cloud did not surrender, die young, or flee his country into exile. He stayed with his people and did his best to lead them through trying times. During the past half-century, however, Red Cloud's reputation has suffered. Meanwhile, Crazy Horse has become the great symbol of Indian resistance, his resolution and tragic death symbolized by a monumental granite carving near Custer, South Dakota—the town named for his famous enemy. The Battle of the Little Bighorn has become the idealized epitome of Sioux resistance to subjugation, and while Crazy Horse was there, Red Cloud was not. Furthermore, the taking of the Black Hills remains a prime source of disaffection, and Red Cloud was one of those chiefs who, though bowing to the inevitable, signed the bogus agreement that gave them away. The United States Indian Claims Commission's 1974 award of $17.5 million (plus interest accrued since 1877), which was upheld by the Supreme Court in 1980, has turned out to be a constant source of friction. The Lakota tribes refuse to take the money, now estimated at about $1.4 billion due to accumulated interest, and still press for the return of the land.[22] Perhaps one day, as Red Cloud's role as a fearless advocate for his people during the reservation years becomes better understood, he will secure his place as an Oglala legend. Greatness exceeds implacability.

Notes

PREFACE

1. John D. McDermott, "A Dedication to the Memory of George Hyde, 1882–1968," *Arizona and the West* 17 (Summer 1975): 104.

2. George E. Hyde, *Red Cloud's Folk: A History of the Oglala Sioux Indians*, 1st ed. (Norman: University of Oklahoma Press, 1937), *A Sioux Chronicle* (Norman: University of Oklahoma Press, 1956), and *Spotted Tail's Folk: A History of the Brulé Sioux* (Norman: University of Oklahoma Press, 1961). A revised edition of *Red Cloud's Folk* was published in 1975.

3. James C. Olson, *Red Cloud and the Sioux Problem* (Lincoln: University of Nebraska Press, 1965), p. vii.

4. Ibid., p. viii.

5. R. Eli Paul, "Recovering Red Cloud's Autobiography: The Strange Odyssey of a Chief's Personal Narrative," *Montana, the Magazine of Western History* 44 (Summer 1994): 2–17.

6. R. Eli Paul, ed., *Autobiography of Red Cloud, War Leader of the Oglalas* (Helena: Montana Historical Society Press, 1997), pp. 8–23.

7. Robert W. Larson, *Red Cloud: Warrior-Statesman of the Lakota Sioux* (Norman: University of Oklahoma Press, 1997), p. xiv.

8. Ibid., p. 304.

9. Catherine Price, *The Oglala People, 1841–1879: A Political History* (Lincoln: University of Nebraska Press, 1996); Ed McGaa, *Crazy Horse and Chief Red Cloud: Warrior Chiefs–Teton Oglalas* (Minneapolis: Four Directions Publishing, 2005); Kingsley M. Bray, *Crazy Horse: A Lakota Life* (Norman: University of Oklahoma Press, 2006); Thomas Powers, *The Killing of Crazy Horse* (New York: Alfred A. Knopf, 2010).

10. John D. McDermott, *Circle of Fire: The Indian War of 1865* (Mechanicsburg, Pa.: Stackpole Books, 2003), and *Red Cloud's War: The Bozeman Trail, 1866–1868*, 2 vols. (Norman, Okla.: Arthur H. Clark, 2010).

11. Doane Robinson, "The Education of Red Cloud," *South Dakota Historical Collections* 12 (1924): 156–78, and "The Sioux of the Dakotas," *Home Geographic Monthly* 2 (Nov. 1932): 8–12.

12. Ella Deloria, "Warrior Training," Dakota Ethnology Box, Ella Deloria Collection, Dakota Indian Foundation, Chamberlain, S.Dak.

13. For the *Council Fire*'s publication history, *see* Jo Lea Wetherilt Behrens, "In Defense of 'Poor Lo': National Indian Defense Association and *Council Fire*'s Advocacy for Sioux Land Rights," *South Dakota History* 24 (Fall/Winter 1994): 153–73.

CHAPTER 1: RED CLOUD'S WORLD

1. Leslie Thompson Dykstra, "Red Cloud," *Nebraska History Magazine* 15 (Jan.–Mar. 1934): 3.

2. The other large groupings were the Sissetons, Wahpetons, Wahpekutes, Mdewakantons, Yanktons, and Yanktonais. The first four, also known as the Santees, called themselves the Dakotas, meaning "friends" or "allies," which they preferred to "Sioux," the French corruption of a name given them by the Ojibways. Because of dialectical differences, the Tetons became the Lakotas, and the Yanktons and Yantonais were often called the Nakotas.

3. Stephen R. Riggs, *Dakota Grammar, Texts, and Ethnography*, ed. James Owen Dorsey (1893; reprint, Marvin, S.Dak.: American Indian Culture Research Center, 1977), p. 163.

4. Horses, introduced into southwestern North America by the Spanish in the seventeenth century, diffused gradually across the continent via intertribal trade networks. The Sioux most likely acquired horses in the eighteenth century. For the latest scholarship on the Teton Sioux, *see* Raymond J. DeMallie, "Sioux until 1850" and "Teton," in *Handbook of North American Indians*, vol. 13, ed. William C. Sturtevant (Washington, D.C.: Smithsonian Institution, 2001), pp. 718–60, 794–820.

5. W. W. Newcomb, Jr., "A Re-examination of the Causes of Plains Warfare," *American Anthropologist* 52 (July-Sept. 1950): 321, 325; James H. Cook, *Fifty Years on the Old Frontier as Cowboy, Hunter, Guide, Scout, and Ranchman* (Norman: University of Oklahoma Press, 1957), p. 187.

6. American Horse's family recorded in its winter count, "Standing Bear led the first party of Oglalas to the Black Hills in 1775 or 1776" (quoted in Collin G. Calloway, *One Vast Winter Count: The Native American West before Lewis and Clark* [Lincoln: University of Nebraska Press, 2003], p. 309). *See also* Linea Sundstrom, *Storied*

Stone: Indian Rock Art in the Black Hills Country (Norman: University of Oklahoma Press, 2004), p. 18.

7. Kingsley M. Bray, "The Oglala Lakota and the Establishment of Fort Laramie," *Museum of the Fur Trade Quarterly* 36 (Winter 2000):14.

8. Red Hawk, quoted in James R. Walker, *Lakota Belief and Ritual*, ed. Raymond J. DeMallie and Elaine A. Jahner (Lincoln: University of Nebraska Press, 1980), p. 137.

9. R. Eli Paul, ed., *Autobiography of Red Cloud, War Leader of the Oglalas* (Helena: Montana Historical Society Press, 1997), p. 2; Joseph Agonito, *Lakota Portraits: Lives of the Legendary Plains People* (Guilford, Conn.: Twodot, 2011), pp. 25–26.

10. Luther Standing Bear, *Land of the Spotted Eagle* (Lincoln: University of Nebraska Press, 1978), p. 193.

11. Raymond J. DeMallie, "Lakota Belief and Ritual in the Nineteenth Century," in *Sioux Indian Religion: Tradition and Innovation*, ed. DeMallie and Douglas R. Parks (Norman: University of Oklahoma Press, 1987), p. 28.

12. Red Cloud, quoted in Walker, *Lakota Belief and Ritual*, p. 140.

13. Christopher Vecsey, "American Indian Environmental Religions," in *American Indian Environments: Ecological Issues in Native American History*, ed. Vecsey and Robert W. Venables (Syracuse, N.Y.: Syracuse University Press, 1980), p. 19.

14. John D. McDermott, *A Guide to the Indian Wars of the West* (Lincoln: University of Nebraska Press, 1998), pp. 1–2; Raymond J. DeMallie, "Touching the Pen: Plains Indian Treaty Councils in Ethnohistorical Perspective," in *Ethnicity on the Great Plains*, ed. Frederick C. Luebke (Lincoln: University of Nebraska Press, 1980), p. 47; George Bird Grinnell, "Tenure of Land among the Indians," *American Anthropologist* 9 (Jan.–May 1907): 1–2.

15. Red Cloud, quoted in Walker, *Lakota Belief and Ritual*, p. 138. *See also* William T. Hagan, "Justifying Dispossession of the Indian," in *American Indian Environments*, ed. Vecsey and Venables, p. 65ff.

16. Lawrence A. Frost, ed., *With Custer in '74: James Calhoun's Diary of the Black Hills Expedition* (Provo, Utah: Brigham Young University Press, 1979), p. 40.

17. Red Cloud, quoted in Walker, *Lakota Belief and Ritual*, p.138.

CHAPTER 2: PREPARATION FOR GREATNESS

1. James H. Cook, *Fifty Years on the Old Frontier as Cowboy, Hunter, Guide, Scout, and Ranchman* (Norman: University of Oklahoma Press, 1957), p. 202.

2. Red Cloud, quoted in Doane Robinson, "The Sioux of the Dakotas," *Home Geographic Monthly* 2 (Nov. 1932): 7–12. *See also* Robinson, "The Education of Red Cloud," *South Dakota Historical Collections* 12 (1924): 156–78. It should be remembered, however, that Red Cloud almost certainly spoke to Robinson through an interpreter.

3. Charles A. Eastman, *Indian Heroes and Great Chieftains* (Lincoln: University of Nebraska Press, 1991), p. 4.

4. Ibid., pp. 4–6.

5. Royal B. Hassrick, *The Sioux: Life and Customs of a Warrior Society* (Norman: University of Oklahoma Press, 1964), p. 32.

6. Stephen R. Riggs, *Dakota Grammar, Texts, and Ethnography*, ed. James Owen Dorsey (1893; reprint, Marvin, S.Dak.: American Indian Culture Research Center, 1977), p. 209.

7. Ella Deloria, "Warrior Training," pp. 133–35, Dakota Ethnology Box, Ella Deloria Collection, Dakota Indian Foundation, Chamberlain, S.Dak.; Hassrick, *The Sioux*, p. 73.

8. Deloria, "Warrior Training," pp. 133–35.

9. Edward S. Curtis, *The North American Indian*, vol. 3 (Norwood, Mass.: By the Author, 1909), p. 187.

10. Cook, *Fifty Years*, p. 202.

11. Ibid.; James H. Cook to Charles G. Coutant, 20 Mar. 1898, Folder 39, Coutant Collection, Wyoming State Archives, Cheyenne; Valentine T. McGillycuddy to Walter M. Camp, 5 Dec. 1919, Walter M. Camp Papers, Little Bighorn Battlefield National Monument, Crow Agency, Mont.; Charles W. Allen, "Red Cloud, Chief of the Sioux," *The Hesperian* 1 (Jan. 1896): 211–16; Edmund B. Tuttle, *Three Years on the Plains: Observations of Indians, 1867–1870* (Norman: University of Oklahoma Press, 2003), p. 101.

12. James H. Cook, "The Art of Fighting Indians," *American Mercury* 23 (June 1931): 171.

13. Thisba Hutson Morgan, "Reminiscence of My Days in the Land of the Oglala Sioux," *South Dakota Historical Collections* 29 (1958): 32.

14. Red Cloud war songs collected from Jim Red Cloud, Kills Above, and Knife Chief by James H. Cook and translated by Joe Si-

erro, 1931, James H. Cook Papers, Agate Fossil Beds National Monument, Harrison, Nebr.

15. Eastman, *Indian Heroes*, pp. 6–7.

16. Kingsley M. Bray, "The Oglala Lakota and the Establishment of Fort Laramie," *Museum of the Fur Trade Quarterly* 36 (Winter 2000): 10–12.

17. Peter Richard married Red Sack's daughter, the granddaughter of No Water. *See* interview with Frank Aplan, 17 Sept. 1963, and "Writings of Alice Brown," copies in John D. McDermott Collection, Rapid City, S.Dak.; Statement of Julia Richard, Julia Richard Claim #4565, Records of the United States Court of Claims, Record Group 123, National Archives, Washington, D.C.; Robinson, "Sioux of the Dakotas," p. 11. For a history of the Richard family, *see* John D. McDermott, "John Baptiste Richard," *The Mountain Men and the Fur Trade of the Far West*, vol. 2, ed. LeRoy R. Hafen (Glendale, Calif.: Arthur H. Clark, 1965), pp. 289–303.

18. Red Cloud, quoted in Robinson, "Sioux of the Dakotas," p. 12.

19. Robert M. Utley, "The Bozeman Trail before John Bozeman: A Busy Land," *Montana, the Magazine of Western History* 53 (Summer 2003): 21–22.

20. Cook, *Fifty Years*, p. 202.

21. Valentine T. McGillycuddy to Elmo Scott Watson, 10 Feb. 1922, Elmo Scott Watson Papers, Newberry Library, Chicago, Ill.; George E. Hyde, *Red Cloud's Folk: A History of the Oglala Sioux Indians*, rev. ed. (Norman: University of Oklahoma Press, 1975), pp. 34n9, 305.

22. R. Eli Paul, ed., *Autobiography of Red Cloud, War Leader of the Oglalas* (Helena: Montana Historical Society Press, 1997), pp. 34–35.

23. Hyde, *Red Cloud's Folk*, p. 305.

CHAPTER 3: RISE TO POWER

1. American Horse, interview by Eli S. Ricker, in *Voices of the American West: The Indian Interviews of Eli S. Ricker, 1903–1919*, ed. Richard E. Jensen (Lincoln: University of Nebraska Press, 2005), p. 281.

2. Clarence Three Stars stated that the name "Bad Face" came about because during a hard winter, marked by deep snow and depleted game, the band did not hunt buffalo. Instead, they hung

around the lodges of other Lakotas, depending on their help to survive. Consequently, the people's faces "hung down" in unhappiness, causing them to be called Bad Faces. Clarence Three Stars, interview by Eli S. Ricker, in *Indian Interviews*, ed. Jensen, p. 348.

3. Eagle Hawk, interview by Addison E. Sheldon, 28 July 1903, Box 59, Addison E. Sheldon Collection, Nebraska State Historical Society, Lincoln; R. Eli Paul, ed., *Autobiography of Red Cloud, War Leader of the Oglalas* (Helena: Montana Historical Society Press, 1997), pp. 64–70; James C. Olson, *Red Cloud and the Sioux Problem* (Lincoln: University of Nebraska Press, 1965), p. 20.

4. Paul, *Autobiography of Red Cloud*, pp. 64–65, 67–70. *See also* Charles W. Allen, "Red Cloud, Chief of the Sioux," *The Hesperian* 1 (Jan. 1896): 211–16.

5. George E. Hyde, *Red Cloud's Folk: A History of the Oglala Sioux Indians*, rev. ed. (Norman: University of Oklahoma Press, 1975), p. 53. *See also* Robert W. Larson, *Red Cloud: Warrior-Statesman of the Lakota Sioux* (Norman: University of Oklahoma Press, 1997), pp. 58–59.

6. Olson, *Red Cloud*, p. 22; Larson, *Red Cloud*, p. 60; Paul, *Autobiography of Red Cloud*, p. 65.

7. Mari Sandoz, *Crazy Horse: The Strange Man of the Oglalas*, 3d ed. (Lincoln: University of Nebraska Press, 2008), pp. 444, 462, 470–71; Little Wound, quoted in James R. Walker, *Lakota Belief and Ritual*, ed. Raymond J. DeMallie and Elaine A. Jahner (Lincoln: University of Nebraska Press, 1980), p. 195.

8. Raymond J. DeMallie, "Male and Female in Traditional Lakota Culture," in *The Hidden Half: Studies of Plains Indian Women*, ed. Patricia Albers and Beatrice Medicine (Lanham, Md.: University Press of America, 1983), p. 252.

9. Paul, *Autobiography of Red Cloud*, p. 76.

10. Ibid., pp. 76–82.

11. Ibid., pp. 76–84.

12. Nick Janis, quoted in Willis Fletcher Johnson, *Life of Sitting Bull and History of the Indian War of 1890–91* (Edgewood, S.Dak.: Edgewood Publishing Co., 1891), p. 111. Johnson uses the variant spelling "Genneiss" in his work.

13. Janis, quoted ibid., pp. 108–9.

14. Susan Sleeper-Smith, "Furs and Female Kin Networks: The World of Marie Madeleine Réaume L'archeveque Chevalier," in *New*

Faces of the Fur Trade: Selected Papers of the Seventh North American Fur Trade Conference, Halifax, Nova Scotia, 1995, ed. Jo-Anne Fiske, Susan Sleeper-Smith, and William Wicken (Lansing: Michigan State University Press, 1998), pp. 53–55, 63; Marla N. Powers, Oglala Women: Myth, Ritual, and Reality (Chicago: University of Chicago Press, 1986), pp. 7–8.

15. Paul, Autobiography of Red Cloud, pp. 106–7; Larson, Red Cloud, p. 44; U.S., Department of the Interior, Office of Indian Affairs, "Census of Indians at Pine Ridge Agency, D.T.: Ogallala Sioux and Mixed Bloods," [1887], p. 4, in Indian Census Rolls, 1885–1940, National Archives Microfilm Publication M595 (Washington, D.C.: National Archives, 1965), Roll 363; George Bird Grinnell, "Red Cloud," ca. 1891, Folder 10, Box 2, Series 3, George Bird Grinnell Collection MO610, Nebraska State Historical Society, Lincoln; Pittsburg Dispatch, 8 July 1892.

16. Paul, Autobiography of Red Cloud, pp. 106–11; Ella Deloria, "Warrior Training," pp. 133–34, Dakota Ethnology Box, Ella Deloria Collection, Dakota Indian Foundation, Chamberlain, S.Dak.

CHAPTER 4: ROAD TO WAR

1. Red Cloud, quoted in George Bird Grinnell, "Red Cloud," c. 1891, Folder 10, Box 2, Series 3, George Bird Grinnell Collection MO610, Nebraska State Historical Society, Lincoln.

2. Jeffrey Ostler, The Plains Sioux and U.S. Colonialism from Lewis and Clark to Wounded Knee (Cambridge, U.K.: Cambridge University Press, 2004), p. 33. See also Esther S. Goldfrank, "Historic Change and Social Character: A Study of the Teton Dakota," American Anthropologist 45 (Jan.–Mar. 1943): 74.

3. "Speech of Red Cloud," unidentified newspaper clipping, c. June 1870, Individual Indian Speeches, Box 5, Folder 14, James Brisbin Collection, Montana Historical Society, Helena.

4. Ostler, Plains Sioux, pp. 36–37; Catherine Price, The Oglala People, 1841–1879: A Political History (Lincoln: University of Nebraska Press, 1996), pp. 30–35. See also Charles J. Kappler, ed., Indian Treaties, 1778–1883, vol. 2 (1904; reprint, New York: Interland Publishing, 1972), p. 595.

5. Edwin T. Denig, Five Indian Tribes of the Upper Missouri: Sioux, Arickaras, Assiniboines, Crees, Crows, ed. John C. Ewers (Norman: University of Oklahoma Press, 1969), p. 139.

6. Henry B. Carrington to H. G. Litchfield, 26 Sept. 1866, in *Indian Operations on the Plains*, Senate Ex. Doc. No. 33, 50th Cong., 1st sess., 1887, p. 27; Colin G. Calloway, "The Only Way Open to Us: The Crow Struggle for Survival in the Nineteenth Century," *North Dakota History* 33 (Summer 1980): 27.

7. R. Eli Paul, *Blue Water Creek and the First Sioux War, 1854–1856* (Norman: University of Oklahoma Press, 2004), pp. 15–16; Price, *Oglala People*, pp. 37–38.

8. Frank Salaway, interview by Eli S. Ricker, in *Voices of the American West: The Indian Interviews of Eli S. Ricker, 1903–1919,* ed. Richard E. Jensen (Lincoln: University of Nebraska Press, 2005), pp. 332–35; Price, *Oglala People*, pp. 37–40; Robert W. Larson, *Red Cloud: Warrior-Statesman of the Lakota Sioux* (Norman: University of Oklahoma Press, 1997), pp. 67–70; Doane Robinson, "The Education of Red Cloud," *South Dakota Historical Collections* 12 (1924): 164.

9. Ostler, *Plains Sioux*, pp. 41–42; Price, *Oglala People*, pp. 41–43.

10. Ostler, *Plains Sioux*, p. 42.

11. Kingsley M. Bray, "Lone Horn's Peace: A New View of Sioux-Crow Relations, 1851–1858," *Nebraska History* 66 (1985): 39, 44; R. Eli Paul, ed., *Autobiography of Red Cloud, War Leader of the Oglalas* (Helena: Montana Historical Society Press, 1997), pp. 118–24, 177–87.

12. Robert M. Utley, "The Bozeman Trail before John Bozeman: A Busy Land," *Montana, the Magazine of Western History* 53 (Summer 2003): 23; Bray, "Lone Horn's Peace," pp. 44–45.

13. Margaret Irvin Carrington, *Absaraka, Home of the Crows* (Philadelphia: J. B. Lippincott, 1868), p. 17.

14. Price, *Oglala People*, p. 45.

15. Ibid.; John D. McDermott, *Circle of Fire: The Indian War of 1865* (Mechanicsburg, Pa.: Stackpole Books, 2003), pp. 3–6.

16. Ostler, *Plains Sioux*, p. 44; Price, *Oglala People*, p. 53.

17. Quoted in "What the Indians of the Northwest Thought" (clipping), *Billings Gazette*, n.d., in "Indian Scrapbook," pp. 8–9, Billings Public Library, Billings, Mont. See also Charles A. Eastman, *Indian Heroes and Great Chieftains* (Lincoln: University of Nebraska Press, 1991), p. 15.

18. Raymond J. DeMallie and Douglas A. Parks, "Plains Indian Warfare," in *The People of the Buffalo: Plains Indians of North America, Military Art, Warfare and Change*, ed. Colin F. Taylor and Hugh A. Dempsey (Wyk auf Foehr, Germany: Tatanka Press, 2003),

p. 72; Charles Rambow, *Bear Butte: Journeys to the Sacred Mountain* (Sioux Falls, S.Dak.: Pine Hill Press, 2004), pp. 35–36.

19. McDermott, *Circle of Fire*, pp. 12–15.

20. Ibid., pp. 13–14.

21. Ibid., pp. 15–34.

22. John D. McDermott, "'We had a terribly hard time letting them go': The Battles of Mud Springs & Rush Creek, February, 1865," *Nebraska History* 77 (Summer 1996): 78–88.

23. For a detailed account of the battles, *see* John D. McDermott, *Frontier Crossroads: A History of Fort Caspar and the Upper Platte Crossing* (Casper, Wyo.: City of Casper, 1999), pp. 61–74. Two earlier treatments are J. W. Vaughn, *The Battle of Platte Bridge* (Norman: University of Oklahoma Press, 1963), and Alfred J. Mokler, *Fort Caspar (Platte Bridge Station)* (Casper, Wyo.: Prairie Publishing Co., 1939). *See also* McDermott, *Circle of Fire*, p. 86.

24. McDermott, *Circle of Fire*, pp. 86–99; David Fridtjof Halaas and Andrew E. Masich, *Halfbreed: The Remarkable True Story of George Bent—Caught Between the Worlds of the Indian and the White Man* (Cambridge, Mass.: Da Capo Press, 2004), pp. 184–85; Verne Mokler, "Story of Caspar Collins: Oregon Trail Trek No. Three," *Annals of Wyoming* 28 (Oct. 1956): 186; Wayne Sundberg, *The Story of Camp and Fort Collins* (Fort Collins, Colo.: n.p., 1985), pp. 1–3; Agnes Wright Spring, "The Founding of Fort Collins, United States Military Post," *Colorado Magazine* 10 (1933): 47–55. *See also* McDermott, *Frontier Crossroads*, pp. 66–82.

25. John D. McDermott, *A Guide to the Indian Wars of the West* (Lincoln: University of Nebraska Press, 1998), pp. 48–52.

26. John D. McDermott, *Red Cloud's War: The Bozeman Trail, 1866–1868*, 2 vols. (Norman, Okla.: Arthur H. Clark, 2010), 1:17; McDermott, *Circle of Fire*, pp. 8, 100–19.

27. McDermott, *Circle of Fire*, pp. 120–28.

28. Ibid., pp. 100–44.

29. Reginald and Gladys Laubin, *The Indian Tipi: Its History, Construction and Use* (New York: Ballantine Books, 1957), p. 211.

CHAPTER 5: NEGOTIATIONS FAIL

1. Remington Schuyler, quoted in *Denver Farm and Fireside*, Jan. 1922, copy in Elmo Scott Watson Papers, Newberry Library, Chicago, Ill.

2. John D. McDermott, *Circle of Fire: The Indian War of 1865* (Mechanicsburg, Pa.: Stackpole Books, 2003), pp. 146–47; Robert W. Larson, *Red Cloud: Warrior-Statesman of the Lakota Sioux* (Norman: University of Oklahoma Press, 1997), p. 89; James C. Olson, *Red Cloud and the Sioux Problem* (Lincoln: University of Nebraska Press, 1965), p. 28.

3. For a detailed account of the funeral, *see* McDermott, *Circle of Fire*, pp. 153–57.

4. Quoted in Olson, *Red Cloud*, p. 32.

5. Ibid., p. 31; John D. McDermott, *Red Cloud's War: The Bozeman Trail, 1866–1868*, 2 vols. (Norman, Okla.: Arthur H. Clark, 2010), 1:35.

6. Edward B. Taylor to Red Cloud (telegram), Fort Laramie, 12 Mar. 1866, Letters Received, Records of the Upper Platte Agency, Records of the Office of Indian Affairs, Record Group 75, National Archives, Washington, D.C. (hereafter cited as Upper Platte Agency Records). *See also* McDermott, *Red Cloud's War*, 1:35–36.

7. Olson, *Red Cloud*, pp. 31–33.

8. McDermott, *Red Cloud's War*, p. 51.

9. William F. Hynes, *Soldiers of the Frontier* (N.p.: Grand Council Fire of American Indians, 1943), p. 44. *See also* McDermott, *Red Cloud's War*, pp. 51–52.

10. McDermott, *Red Cloud's War*, p. 53.

11. Edward B. Taylor, quoted in *Denver Daily Rocky Mountain News,* 18 June 1866. *See also* McDermott, *Red Cloud's War*, 1:53.

12. Red Cloud, quoted in *Denver Daily Rocky Mountain News*, 18 June 1866.

13. Red Cloud, quoted ibid.

14. Edward B. Taylor to Commissioner of Indian Affairs, 9 June 1866, Upper Platte Agency Records.

15. *Salt Lake City Union Vedette*, 22 June 1866.

16. Olson, *Red Cloud*, pp. 34–35; McDermott, *Red Cloud's War*, 1:54–55.

17. McDermott, *Red Cloud's War*, 1:56–57.

18. John Monnett, *Where a Hundred Soldiers Were Killed: The Struggle for the Powder River Country in 1868 and the Making of the Fetterman Myth* (Albuquerque: University of New Mexico Press, 2008), pp. 27, 41; *Denver Daily Rocky Mountain News*, 25 June 1866; *Salt Lake City Union Vedette*, 29 June 1866. Kingsley M. Bray states

that Red Cloud and Man Afraid of His Horses returned to Fort Laramie on 12 June to meet Spotted Tail, among other Lakota leaders,
and lobby against participation in the peace council. *See* Bray, *Crazy
Horse: A Lakota Life* (Norman: University of Oklahoma Press, 2006),
pp. 92–93.

19. Quoted in Margaret Irvin Carrington, *Absaraka, Home of the
Crows* (Philadelphia: J. B. Lippincott, 1868), p. 185.

CHAPTER 6: RED CLOUD'S WAR

1. Henry B. Carrington, quoted in Margaret I. and Henry B. Carrington, *Absaraka, Land of Massacre*, 3d ed. (Philadelphia: J. B. Lippincott, 1878), p. 266.

2. "From Julesburg," *Salt Lake Daily Union Vedette*, 1 Mar. 1867.

3. John D. McDermott, *Red Cloud's War: The Bozeman Trail,
1866–1868*, 2 vols. (Norman, Okla.: Arthur H. Clark, 2010), 1:76. *See
also* McDermott, *A Guide to the Indian Wars of the West* (Lincoln: University of Nebraska Press, 1998), pp. 51–52.

4. McDermott, *Red Cloud's War*, 1:92, 201–2; Henry B. Carring-
ton, "Monument Dedication Speech," in *My Army Life and the Fort
Phil Kearney Massacre*, by Frances C. Carrington (Philadelphia: J. B.
Lippincott, 1910), p. 254.

5. Carrington, *My Army Life*, p. 93; Francis Hynes, *Soldiers of the
Frontier* (N.p.: Grand Council Fire of American Indians, 1943), p. 60.

6. McDermott, *Guide to the Indian Wars*, p. 52, *Red Cloud's War*,
1:201–29, and "The Short and Controversial Life of William Judd
Fetterman," *Annals of Wyoming* 63 (Spring 1991): 49–53; James H.
Cook, *Fifty Years on the Old Frontier as Cowboy, Hunter, Guide, Scout,
and Ranchman* (Norman: University of Oklahoma Press, 1957), p.
198.

7. McDermott, *Red Cloud's War*, 1:202–3, 203n11; Robert W. Larson, *Red Cloud: Warrior-Statesman of the Lakota Sioux* (Norman: University of Oklahoma Press, 1997), pp. 99–104; Catherine Price, *The
Oglala People, 1841–1879: A Political History* (Lincoln: University of
Nebraska Press, 1996), p. 67.

8. McDermott, *Red Cloud's War*, 2:407–37.

9. Red Cloud, quoted in *Omaha World Herald*, 9 Aug. 1903.

10. McDermott, *Red Cloud's War*, 2:379–405.

11. Ibid., 1:xiii, 2:439–40; Susan Badger Doyle, "The Bozeman
Trail, 1863–1868," *Overland Journal* 20 (Spring 2002): 11.

12. Larson, *Red Cloud*, p. 94; McDermott, *Red Cloud's War*, 1:127, 2:453–59.

13. "Report of the Peace Commissioners, January 7, 1868," in *Taopi and His Friends or the Indians' Wrongs and Rights* (Philadelphia: Claxton, Remsen & Haffelfinger, 1869), pp. 89, 96–97.

14. McDermott, *Red Cloud's War*, 2:483–85, 522–25.

15. Ibid., 2:484; Adam J. Slemmer to Assistant Adjutant General, Department of the Platte, 5 Mar. 1868, Letters Sent, Fort Laramie Post Records, Records of United States Army Commands, Record Group 393, National Archives, Washington, D.C.

16. John S. Gray, *Custer's Last Campaign: Mitch Boyer and the Little Bighorn Reconstructed* (Lincoln: University of Nebraska Press, 1991), p. 79; McDermott, *Red Cloud's War*, 2:495–97. For more on Spotted Tail's role, *see* George E. Hyde, *Spotted Tail's Folk: A History of the Brulé Sioux* (Norman: University of Oklahoma Press, 1961), pp. 110–26.

17. Quoted in McDermott, *Red Cloud's War*, 2:497–98. For the full text of the treaty, *see* Charles J. Kappler, ed., *Indian Affairs: Laws and Treaties* (Washington, D.C.: Government Printing Office, 1904), 2:989–1007. The quotations are from articles II and XI.

18. McDermott, *Red Cloud's War*, 2:497–98.

19. Ibid., p. 498; Jeffrey Ostler, *The Plains Sioux and U.S. Colonialism from Lewis and Clark to Wounded Knee* (Cambridge, U.K.: Cambridge University Press, 2004), pp. 49–50.

20. McDermott, *Red Cloud's War*, 2:501–5.

21. *Cheyenne Leader*, 6 June 1868.

22. Slemmer to Assistant Adjutant General, Department of the Platte, 5 June 1868, Fort Laramie Post Records.

23. McDermott, *Red Cloud's War*, 2:522–27.

24. Ibid., 2:534.

25. Robert W. Larson, "Lakota Leaders and Government Agents: A Story of Changing Relationships," *Nebraska History* 82 (Summer 2001): 48. *See also* Garrick Mallery, *Picture-Writing of the American Indians*, 2 vols. (New York: Dover Publications, 1972), 2:650.

26. McDermott, *Red Cloud's War*, 2:534–35.

27. Ibid., p. 535.

28. Ann Vogdes, Journal, 5 Nov. 1868, quoted in Sherry L. Smith, "Officers' Wives, Indians, and the Indian Wars," *Order of the Indian Wars Journal* 1 (Winter 1980): 35.

29. McDermott, *Red Cloud's War*, 2:535–36.

30. Ibid., p. 536.

31. William M. Dye to George D. Ruggles, 5 Dec. 1868, Fort Laramie Post Records.

32. Price, *Oglala People*, pp. 118–19.

33. Raymond J. DeMallie, *The Sixth Grandfather: Black Elk's Teachings Given to John G. Neihardt* (Lincoln: University of Nebraska Press, 1984), p. 323; Larson, *Red Cloud*, pp. 124–25; Price, *Oglala People*, p. 83; James McLaughlin, *My Friend the Indian* (Boston: Houghton Mifflin, 1910), pp. 20–21; Ostler, *Plains Sioux*, p. 51.

34. McDermott, *Red Cloud's War*, 2: 537–38; Thomas Powers, *The Killing of Crazy Horse* (New York: Alfred A. Knopf, 2010), pp. 27–29.

35. John R. Brennan, "Old Red Cloud the Sioux Chief Dead," *Rapid City Journal*, 11 Dec. 1909.

36. Donald Berthrong, "Changing Concepts: The Indians Learn about the 'Long Knives' and Settlers, 1849–1890s," in *Red Men and Hat Wearers: Viewpoints in Indian History*, ed. Daniel Tyler (Fort Collins: Colorado State University Press, 1976), p. 54.

CHAPTER 7: FINDING A PLACE

1. *Omaha Weekly Herald*, 1 June 1870.

2. William M. Dye to Assistant Adjutant General, Department of the Platte, 3 Mar. 1869, Letters Sent, Fort Laramie Post Records, Records of United States Army Commands, Record Group 393, National Archives, Washington, D.C.

3. Jeffrey Ostler, *The Plains Sioux and U.S. Colonialism from Lewis and Clark to Wounded Knee* (Cambridge, U.K.: Cambridge University Press, 2004), pp. 53–54; Catherine Price, *The Oglala People, 1841–1879: A Political History* (Lincoln: University of Nebraska Press, 1996), pp. 83, 86–88.

4. "Red Cloud at Fort Laramie," *Wyoming Weekly Leader*, 3 Apr. 1869. *See also* Remi Nadeau, *Fort Laramie and the Sioux* (Lincoln: University of Nebraska Press, 1967), pp. 248–49.

5. "Red Cloud at Fort Laramie."

6. James C. Olson, *Red Cloud and the Sioux Problem* (Lincoln: University of Nebraska Press, 1965), p. 86.

7. Robert W. Larson, *Red Cloud: Warrior-Statesman of the Lakota Sioux* (Norman: University of Oklahoma Press, 1997), pp. 128–29; Brian Jones, "John Richard, Jr. and the Killing at Fetterman," *Annals*

of Wyoming 43 (Fall 1971): 237–57. The Richard family maintained that John, Jr., found his wife in bed with the soldier and, after drinking heavily, went after him the next day. *See* Frank Aplan, interview by John D. McDermott, 17 Sept. 1963, John D. McDermott Collection, Rapid City, S.Dak.

8. Larson, *Red Cloud*, pp. 128–30; Price, *Oglala People*, p. 90; Olson, *Red Cloud*, p. 97.

9. Olson, *Red Cloud*, pp. 100–101.

10. Red Cloud, quoted ibid., pp. 105–6.

11. Ibid., p. 107.

12. Red Cloud, quoted ibid., pp. 107–8. *See also* Price, *Oglala People*, p. 91; Larson, *Red Cloud*, p. 133.

13. Olson, *Red Cloud*, p. 108.

14. Ibid., pp. 108–10; Price, *Oglala People*, pp. 91–92; Larson, *Red Cloud*, pp. 133–34.

15. Olson, *Red Cloud*, pp. 110–13; Larson, *Red Cloud*, pp. 134–35.

16. Red Cloud, quoted in Olson, *Red Cloud*, pp. 112–13. Red Cloud's Cooper Union speech has been widely reprinted, sometimes with minor variations from the text given here. For example, *see* Bob Blaisdell, ed., *Great Speeches by Native Americans* (Mineola, N.Y.: Dover Publications, 2000), pp. 132–33.

17. Larson, *Red Cloud*, p. 135.

18. *New York Times*, 8 June 1870, quoted in Olson, *Red Cloud*, p. 126.

19. *New York Herald*, 9 June 1870, quoted ibid., pp. 126–27.

20. George E. Hyde, *Red Cloud's Folk: A History of the Oglala Sioux Indians*, rev. ed. (Norman: University of Oklahoma Press, 1975), p. 181.

21. "Red Cloud Returns," *Cheyenne Daily Leader*, 22 June 1870.

22. Jones, "John Richard, Jr.," pp. 255–56.

23. Larson, *Red Cloud*, pp. 137–38; Olson, *Red Cloud*, p. 131. *See also* Martin Luschei, *The Black Hills and the Indians: Haven of Our Hopes* (San Luis Obispo, Calif.: Niobrara Press, 2007), pp. 43–44.

24. Price, *Oglala People*, pp. 92–95; Olson, *Red Cloud*, pp. 118–19.

25. Price, *Oglala People*, p. 98; Carol A. Ripich, "Joseph W. Wham and the Red Cloud Agency," *Arizona and the West* 12 (Winter 1970): 330, 333. The Grant Administration's "Peace Policy," implemented in 1869, sought to "civilize" American Indians by settling them on reservations and providing them instruction in Euro-American

ways of life. Christian churches were an integral part of the plan, as they would nominate candidates for the office of Indian agent and dispatch missionaries to the reservations. *See* Price, *Oglala People*, pp. 85–86, and Ostler, *Plains Sioux*, pp. 187–88.

26. Red Cloud, quoted in U.S., Office of Indian Affairs, *Annual Report of the Commissioner of Indian Affairs to the Secretary of the Interior for the Year 1871* (Washington, D.C.: Government Printing Office, 1872), p. 22.

27. Price, *Oglala People*, pp. 98–100.

28. Red Cloud, quoted in *Annual Report of the Commissioner*, p. 26.

29. Ibid., pp. 28–29.

30. Price, *Oglala People*, pp. 100–101; Larson, *Red Cloud*, pp. 144–45.

31. Larson, *Red Cloud*, p. 147; Price, *Oglala People*, p. 105; Ripich, "Joseph W. Wham," p. 334.

32. Larson, *Red Cloud*, pp. 147–48; Price, *Oglala People*, pp. 106–8.

33. Price, *Oglala People*, pp. 107–9.

34. Red Cloud, quoted ibid., p. 109.

35. Olson, *Red Cloud*, p. 150.

36. Ibid., pp. 151–52; Price, *Oglala People*, pp. 109–10; Larson, *Red Cloud*, pp. 149–50.

37. Price, *Oglala People*, pp. 111–13.

38. Ibid., pp. 114–20; Olson, *Red Cloud*, pp. 156–58; Larson, *Red Cloud*, pp. 151–52.

39. Robert M. Utley, *The Indian Frontier of the American West, 1846–1890*, rev. ed. (Albuquerque: University of New Mexico Press, 2003), p. 149.

40. Red Cloud, quoted in *Army and Navy Journal*, 20 July 1872, p. 781.

CHAPTER 8: RED CLOUD AGENCY
AND THE BLACK HILLS

1. "Indian Cheek," *Chicago Tribune*, 21 May 1875.

2. James C. Olson, *Red Cloud and the Sioux Problem* (Lincoln: University of Nebraska Press, 1965), pp. 150, 158.

3. U.S., Department of the Interior, Office of Indian Affairs, *Report of the Special Commission Appointed to Investigate the Affairs of*

the Red Cloud Indian Agency, July, 1875, Together with the Testimony and Accompanying Documents (Washington, D.C.: Government Printing Office, 1875), p. xix; Catherine Price, *The Oglala People, 1841–1878: A Political History* (Lincoln: University of Nebraska Press, 1996), pp. 120–21; Olson, *Red Cloud*, pp. 161–63.

4. Price, *Oglala People*, pp. 118–19, 139–41; James A. Walker, *Lakota Society*, ed. Raymond J. DeMallie (Lincoln: University of Nebraska Press, 1982), p. 25.

5. Report of Agent J. J. Saville, Red Cloud Agency, Neb., 31 Aug. 1875, in U.S., Department of the Interior, Office of Indian Affairs, *Annual Report of the Commissioner of Indian Affairs to the Secretary of the Interior for the Year 1875* (Washington, D.C.: Government Printing Office, 1875), pp. 250–51; George E. Hyde, *Red Cloud's Folk*, rev. ed. (Norman: University of Oklahoma Press, 1975), pp. 98, 313. Saville's four groupings match those that anthropologist Clark Wissler described as being in place before the Oglalas moved to the reservation. *See* Wissler, "Societies and Ceremonial Associations in the Oglala Division of the Teton-Dakota," *Anthropological Papers of the American Museum of Natural History* 11 (1912): 7.

6. Report of Saville, p. 250; J. J. Reynolds to Assistant Adjutant General, Department of the Platte, 9 Apr. 1872, Roll 55, Records of Whetstone Agency, Records of the Bureau of Indian Affairs, Record Group 75, National Archives, Washington, D.C.

7. Report of Saville, p. 250; Hyde, *Red Cloud's Folk*, p. 314.

8. Report of Saville, p. 250; Hyde, *Red Cloud's Folk*, p. 193; George E. Hyde, *Spotted Tail's Folk: A History of the Brulé Sioux* (Norman: University of Oklahoma Press, 1961), pp. 247–48.

9. Raymond J. DeMallie, ed., *The Sixth Grandfather: Black Elk's Teachings Given to John G. Neihardt* (Lincoln: University of Nebraska Press, 1984), p. 159n15; "Chief Blue Horse," *Harvard Independent*, 1 Dec. 1910; Hyde, *Red Cloud's Folk*, p. 315.

10. Olson, *Red Cloud*, pp. 164–66; Hyde, *Red Cloud's Folk*, p. 212; Price, *Oglala People*, pp. 128–29.

11. Thomas R. Buecker, *Fort Robinson and the American West, 1874–1899* (Lincoln: Nebraska State Historical Society, 1999), pp. 6–7; Charles W. Allen, "Red Cloud and the U.S. Flag," *Nebraska History* 21 (Oct.-Dec. 1940): 293–304; Price, *Oglala People*, pp. 135–37.

12. *Annual Report of the Commissioner of Indian Affairs . . . 1873*, p. 6; Donald Jackson, *Custer's Gold: The United States Cavalry Expedition of 1874* (New Haven, Conn.: Yale University Press, 1966), p. 15.

13. Valentine T. McGillycuddy, "First Survey of the Black Hills," *Motor Travel* 20 (Oct. 1928): 18.

14. Thomas R. Buecker, "Distance Lends Enchantment to the View: A Diary of the 1874 Black Hills Expedition," in *Beyond Mount Rushmore*, ed. Mary A. Kopco (Pierre: South Dakota State Historical Society Press, 2010), pp. 9–13; Watson Parker, *Gold in the Black Hills* (Norman: University of Oklahoma Press, 1966), p. 8; Ferdinand V. Hayden, "Address on the Black Hills," *Proceedings of the American Philosophical Society* 10 (1869): 322–26.

15. Jackson, *Custer's Gold*, pp. 15–25, 142–44; Olson, *Red Cloud*, pp. 172–74.

16. *New York Tribune*, 10 Aug. 1874; *New York World*, 16 Aug. 1874.

17. *Chicago Inter Ocean*, 1 Aug. 1874.

18. *Bismarck Tribune*, 12 Aug. 1874.

19. *Omaha Weekly Bee*, 8 Apr. 1874.

20. Richard Irving Dodge, *The Black Hills* (1876; reprint, Minneapolis, Minn.: Ross & Haines, 1965), pp. 136–37.

21. Richard I. Dodge, "Report of Black Hills Expedition, May to October, 1875," Reports of Scouts and Reconnaissance, Department of the Platte, 1868–1875, Entry 3817, Records of the Department of the Platte, Record Group 393, National Archives, Washington, D.C.

22. Red Cloud, quoted in *Annual Report of the Commissioner of Indian Affairs . . . 1875*, p. 189. For more on the issue of sacredness, *see* Linea Sundstrom, "The Sacred Black Hills: An Ethnohistorical Review," *Great Plains Quarterly* 17 (Summer/Fall 1997): 206, and Patricia C. Albers et. al., "The Home of the Bison: An Ethnographic and Ethnohistorical Study of Traditional Cultural Affiliations to Wind Cave National Park," contracted study, National Park Service, 2003, pp. 475–529. Sundstrom and Albers have carefully documented the sites and places held sacred by the Sioux prior to 1876. *See also* Red Cloud and Clarence Three Stars, interview by Eli S. Ricker, in *The Indian Interviews of Eli S. Ricker, 1903–1919*, ed. Richard E. Jensen (Lincoln: University of Nebraska Press, 2005), p. 346.

23. Standing Bear, quoted in *Sixth Grandfather*, ed. DeMallie, pp. 163–64.

24. "The Black Hills," *Philadelphia Inquirer*, 24 Aug. 1874; Lawrence A. Frost, ed., *With Custer in '74: James Calhoun's Diary of the Black Hills Expedition* (Provo, Utah: Brigham Young University Press, 1979), p. 41.

25. "The Red Cloud Agency," *Cheyenne Daily Leader*, 3 Apr. 1875.

26. "Spotted Tail at Home," *Omaha Weekly Bee*, 5 May 1875.

27. Grant, quoted in "Indian Cheek," *Chicago Tribune*, 21 May 1875.

28. Ibid.

29. Grant, quoted in "The President and the Sioux," *Cheyenne Daily Leader*, 27 May 1875.

30. Ibid.

31. Olson, *Red Cloud*, p. 201; John D. McDermott, *Red Cloud's War: The Bozeman Trail, 1866–1868*, 2 vols. (Norman, Okla.: Arthur H. Clark, 2010), 1:52; Parker, *Gold in the Black Hills*, pp. 207–8.

32. "Report of the Commission Appointed to Treat with the Sioux Indians for the Relinquishment of the Black Hills" (hereafter cited as Allison Commission Report) in *Report of the Commissioner of Indian Affairs . . . 1875*, p. 186; Olson, *Red Cloud*, p. 205.

33. "The Black Hills," *Omaha Weekly Bee*, 15 Sept. 1875. *See also* Charles Collins, "A Massacre Averted: How Spotted Tail and His Associates Saved the Lives of the Commissioners," *Council Fire* 4 (Sept. 1881): 135, and Martin Luschei, *The Black Hills and the Indians: A Haven of Our Hopes* (San Luis Obispo, Calif.: Niobrara Press, 2007), pp. 80–82.

34. "Life among the Indians," *Omaha Weekly Bee*, 25 Sept. 1875.

35. Olson, *Red Cloud*, p. 206.

36. "Big Indian Council," *Cheyenne Daily Sun-Leader*, 16 Oct. 1897.

37. "The Black Hills," *New York Herald*, 7 Oct. 1875; American Horse, interview by Eli S. Ricker, in *Indian Interviews*, ed. Jensen, p. 283; Allison Commission Report, p. 187; Robert W. Larson, *Red Cloud: Warrior-Statesman of the Lakota Sioux* (Norman: University of Oklahoma Press, 1997), p. 191.

38. Olson, *Red Cloud*, p. 208.

39. Red Cloud, quoted in Allison Commission Report, pp. 188–89.

40. Jeffrey D. Means, "'Indians Shall Do Things in Common': Oglala Lakota Identity and Cattle-Raising," *Montana, the Magazine of Western History* 61 (Autumn 2011): 9.

41. "Red Cloud Agency," *Omaha Daily Bee*, 24 Sept. 1875; "Red Cloud," *Omaha Weekly Bee*, 6,13 Oct., 1875; "The Black Hills," *Chicago Tribune*, 19 Nov. 1875; "Big Indian Council," *Cheyenne Daily Sun-Leader*, 16 Oct. 1897; Hyde, *Red Cloud's Folk*, pp. 244–46; Olson, *Red Cloud*, pp. 208–12; Price, *Oglala People*, pp. 151–52; Larson, *Red Cloud*, pp. 191–93.

42. Allison Commission Report, pp. 194–200. *See also* Olson, *Red Cloud*, pp. 210–13. The *Chicago Tribune* printed the Allison Commission Report in its issue for 18 Nov. 1875.

43. Uri Lanham, *The Bone Hunters: The Heroic Age of Paleontology in the American West* (New York: Dover Publications, 1973), pp. 149–50; Catherine Price, *The Oglala People, 1841–1879: A Political History* (Lincoln: University of Nebraska Press, 1996), p. 142; James C. Olson, *Red Cloud and the Sioux Problem* (Lincoln: University of Nebraska Press, 1965), p. 179; Edward P. Smith to Alexander H. Bullock, 1 July 1875, in U.S., Department of the Interior, *Report of the Special Commission Appointed to Investigate the Affairs of the Red Cloud Indian Agency, July, 1875, Together with the Testimony and Accompanying Documents* (Washington, D.C.: Government Printing Office, 1875), pp. v-vii; John Y. Simon, ed., *The Papers of Ulysses S. Grant*, 32 vols., (Carbondale: Southern Illinois University Press, 1967–2012), 26:xxiv; James J. Saville to Edward P. Smith, 5 June 1875, in *Report of the Special Commission*, pp. iii-iv.

44. Olson, *Red Cloud*, pp. 189–98; "The Investigators," *Omaha Daily Bee*, 27 July 1875; John H. Pierce, "The Indian Ring," *Omaha Daily Bee*, 3 Aug. 1875; "Frauds in Pork," *New York Herald*, 19 Oct. 1875.

45. *Report of the Special Commission*, p. xvii.

46. Olson, *Red Cloud*, p. 216n9; Report of Agent James S. Hastings, 10 Aug. 1876, in U.S., Department of the Interior, Office of Indian Affairs, *Annual Report of the Commissioner of Indian Affairs to the Secretary of the Interior for the Year 1876* (Washington, D.C.: Government Printing Office, 1876), p. 33.

CHAPTER 9: THE GREAT SIOUX WAR
AND ITS RESULTS

1. William T. Sherman in U.S., Department of War, *Annual Report of the General of the Army to the Secretary of War for the Year 1878* (Washington, D.C.: Government Printing Office, 1878), p. 36.

2. Jeffrey Ostler, *The Plains Sioux and U.S. Colonialism from Lewis and Clark to Wounded Knee* (Cambridge, U.K.: Cambridge University Press, 2004), pp. 61–62; Robert W. Larson, *Red Cloud: Warrior-Statesman of the Lakota Sioux* (Norman: University of Oklahoma Press, 1997), p. 198.

3. William H. Jordan to Luther P. Bradley, 24 Apr. 1876, Letters Received, Indian Division, Records of the Secretary of the Interior, Record Group (RG) 48, National Archives (NA), Washington, D.C.

4. Thomas Powers, *The Killing of Crazy Horse* (New York: Alfred A. Knopf, 2010), pp. 166–67.

5. James C. Olson, *Red Cloud and the Sioux Problem* (Lincoln: University of Nebraska Press, 1965), pp. 216–17.

6. Crook, quoted in Olson, *Red Cloud*, p. 217. *See also* "Crook's Campaign," *Cheyenne Daily Leader*, 26 Mar. 1876.

7. "The Indian Campaign," *Baltimore Sun*, 31 July 1876.

8. Larson, *Red Cloud*, pp. 199–204; Olson, *Red Cloud*, p. 223.

9. One Bull, interview by Lewis F. Crawford, Fort Yates, 30 Aug. 1928, Notebook 919, Box 1, Lewis F. Crawford Papers, North Dakota Historical Society, Bismarck; Olson, *Red Cloud*, p. 222.

10. Red Cloud, quoted in Olson, *Red Cloud*, p. 218.

11. "Red Cloud Becomes Civilized," *Indian School Journal* 5 (Feb. 1905): 18–19.

12. Wesley Merritt to Philip H. Sheridan, 5 June 1876, Letters Received, Division of the Missouri, Records of United States Army Commands, RG 393, NA.

13. Seth Bullock to Charles Warren, 8 Sept. 1876, in "From the Black Hills," *Butte* (Mont.) *Miner*, 3 Oct. 1876.

14. "Raiding in the Black Hills," *Butte Miner*, 5 Sept. 1876.

15. "The Indian Inspector's Report," *Chicago Inter-Ocean*, 13 July 1876.

16. Olson, *Red Cloud*, p. 224; Richmond L. Clow, "The Sioux Nation and Indian Territory: The Attempted Removal of 1876," *South Dakota History* 6 (Fall 1976):459–62.

17. "The Sioux Commission," *Chicago Inter-Ocean*, 8 Sept. 1876; "The Great Council," *Cheyenne Daily Leader*, 10 Sept. 1876; Olson, *Red Cloud*, pp. 224–26; Edward Lazarus, *Black Hills, White Justice: The Sioux Nation versus the United States, 1775 to the Present* (New York: HarperCollins, 1991), p. 80.

18. Olson, *Red Cloud*, pp. 226–27. For the full text of the agreement as approved by Congress, *see* Charles J. Kappler, ed., *Indian Affairs: Laws and Treaties*, 5 vols. (Washington, D.C.: Government Printing Office, 1904–1941), 1:168–73.

19. "Indian Intelligence," *St. Paul and Minneapolis Pioneer Press*, 23 Aug. 1876.

20. "The Indian Problem," *New York Herald*, 23 Sept. 1876.

21. Ibid. *See also* Olson, *Red Cloud*, p. 228.

22. George W. Manypenny, "Red Cloud, Chief of the Ogallalla Sioux," *Council Fire* 7 (Dec. 1884):176.

23. Kappler, ed., *Indian Affairs*, 1:170.

24. Olson, *Red Cloud*, pp. 229–31; Lazarus, *Black Hills, White Justice*, p. 93. The issue of removal to Indian Territory is thoroughly discussed in Clow, "Sioux Nation and Indian Territory," pp. 456–73.

25. Olson, *Red Cloud*, pp. 232–34; *Chicago Times*, 15 Nov. 1876.

26. "Visit to Red Cloud's Wigwam," *New York Herald*, 5 Jan. 1877.

27. William Garnett, interview by Eli S. Ricker, in *Voices of the American West: The Indian Interviews of Eli S. Ricker, 1903–1919*, ed. Richard E. Jensen (Lincoln: University of Nebraska Press, 2005), p. 13.

28. George R. Crook to Oliver O. Howard, 20 Feb. 1883, Item 30, George R. Crook Letter Book 1, George R. Crook Papers, Rutherford B. Hayes Presidential Center, Fremont, Ohio.

29. William P. Clark to John G. Bourke, 28 Mar. 1877, Letters Received, Department of the Platte, RG 393, NA.

30. Olson, *Red Cloud*, pp. 237–38; Susan Bordeaux Bettelyoun and Josephine Waggoner, *With My Own Eyes: A Lakota Woman Tells Her People's Story*, ed. Emily Levine (Lincoln: University of Nebraska Press, 1988), p. 109.

31. The best account of this affair can be found in Powers, *Killing of Crazy Horse*, pp. 376–426.

32. Valentine T. McGillycuddy, quoted in John C. Thompson, "In Old Wyoming," *Cheyenne State Tribune*, 29 Nov. 1944.

33. Powers, *Killing of Crazy Horse*, pp. 376–78; Olson, *Red Cloud*, p. 247.

34. Olson, *Red Cloud*, pp. 247–53.

35. Ibid., pp. 252–60.

36. Ibid., pp. 261–63.

1. "Senator Dawes," *Omaha Daily Bee*, 13 Aug. 1884.

2. James N. Leiker and Ramon Powers, *The Northern Cheyenne Exodus in History and Memory* (Norman: University of Oklahoma Press, 2011), pp. 11–13; Candy V. Moulton, *Valentine McGillycuddy: Army Surgeon, Agent to the Sioux* (Norman, Okla.: Arthur H. Clark, 2011), pp. 150–54.

3. James C. Olson, *Red Cloud and the Sioux Problem* (Lincoln: University of Nebraska Press, 1965), p. 264; Valentine T. McGilly-cuddy to Walter M. Camp, 5 Dec. 1919, Walter M. Camp Papers, Little Bighorn Battlefield National Monument, Crow Agency, Mont.

4. Red Cloud to "Our Great Father," 1 May 1879, Entry 71, Pine Ridge Petitions, 1875–1909, Records of the Office of Indian Affairs, Record Group 75, National Archives, Kansas City, Mo.

5. "McGillycuddy-Doctor, Agent, Inventor," *Rapid City Journal*, n. d. [clipping], John D. McDermott Collection, Rapid City, S.Dak.; Moulton, *Valentine McGillycuddy*, pp. 154–57.

6. Moulton, *Valentine McGillycuddy*, pp. 157–58, 229–30. The old standard biography is Julia B. McGillycuddy, *McGillycuddy: Agent* (Stanford, Calif.: Stanford University Press, 1941), now reprinted as Julia B. McGillycuddy, *Blood on the Moon: Valentine McGillycuddy and the Sioux*, intro. James C. Olson (Lincoln: University of Nebraska Press, 1990). Julia McGillycuddy was Valentine McGillycuddy's second wife.

7. Valentine T. McGillycuddy to Elmo Scott Watson, 5 Jan. 1922, Elmo Scott Watson Papers, Newberry Library, Chicago, Ill.

8. Moulton, *Valentine McGillycuddy*, p. 203; Robert W. Larson, *Red Cloud: Warrior-Statesman of the Lakota Sioux* (Norman: University of Oklahoma Press, 1997), pp. 225–27; Jeffrey Ostler, *The Plains Sioux and U.S. Colonialism from Lewis and Clark to Wounded Knee* (Cambridge, U.K.: Cambridge University Press, 2004), pp. 203–5.

9. Red Cloud, quoted in McGillycuddy, *Blood on the Moon*, p. 103.

10. Robert M. Utley, *The Indian Frontier of the American West, 1846–1890* (Albuquerque: University of New Mexico Press, 1984), p. 240; Moulton, *Valentine McGillycuddy*, pp. 160–62.

11. Report of Agent Valentine T. McGillycuddy, Pine Ridge Agency, 1 Sept. 1880, in U.S., Department of the Interior, Office of Indian Affairs, *Annual Report of the Commissioner of Indian Affairs*

to the Secretary of the Interior for the Year 1880 (Washington, D.C.: Government Printing Office, 1880), p. 40.

12. Olson, *Red Cloud*, pp. 288–89.

13. Red Cloud, quoted in George Bird Grinnell, "Red Cloud," ca. 1891, Folder 10, Box 2, Series 3, George Bird Grinnell Collection MO610, Nebraska State Historical Society, Lincoln.

14. Olson, *Red Cloud*, pp. 286–94; Moulton, *Valentine McGilly-cuddy*, pp. 227–28; "The Fraud on the Sioux Exposed," *Council Fire and Arbitrator* 6 (Oct. 1883): 138.

15. Ostler, *Plains Sioux*, p. 204; George E. Hyde, *A Sioux Chronicle* (Norman: University of Oklahoma Press, 1956), pp. 73–74.

16. McGillycuddy, quoted in "Why They Sustain McGillycuddy," *Council Fire and Arbitrator* 7 (Oct. 1884): 147.

17. Moulton, *Valentine McGillycuddy*, pp. 201–2; Doane Robinson, *A History of the Dakota or Sioux Indians* (Minneapolis: Ross & Haines, 1956), pp. 452–54; Ostler, *Plains Sioux*, pp. 155–56; Olson, *Red Cloud*, p. 270.

18. Karl Markus Kreis, ed., *Lakotas, Black Robes, and Holy Women: German Reports from the Indian Missions in South Dakota, 1886–1900*, trans. Corinna Dally-Starna (Lincoln: University of Nebraska Press, 2000), p. 23; Olson, *Red Cloud*, pp. 252, 266–68, 308; Ostler, *Plains Sioux*, p. 188. Red Cloud and his family had reportedly been baptized by Catholic missionaries between 1883 and 1885. *See Lakotas, Black Robes, and Holy Women*, p. 27.

19. Report of Agent Valentine T. McGillycuddy, Pine Ridge Agency, 1 Sept. 1884, in *Report of the Commissioner of Indian Affairs . . . 1884*, p. 37.

20. U.S., Department of the Interior, Office of Indian Affairs, *Rules Governing the Court of Indian Offenses* (Washington, D.C.: Government Printing Office, 1883), pp. 1–8; Ostler, *Plains Sioux*, pp. 175–76, 179.

21. Utley, *Indian Frontier*, p. 243.

22. McGillycuddy, quoted in William T. Hagan, *Indian Police and Judges: Experiments in Acculturation and Control* (Lincoln: University of Nebraska Press, 1966), p. 94.

23. Carrie E. Garrow and Sarah Deer, *Tribal Criminal Law and Procedure* (Walnut Creek, Calif.: AltaMira Press, 2004), p. 87; Hagan, *Indian Police and Judges*, p. 145.

24. Red Cloud et al. to Secretary of the Interior, 13 Aug. 1882, in "The Sioux," *Salt Lake Daily Herald*, 27 Aug. 1882. *See also* Ostler, *Plains Sioux*, p. 205.

25. "Pine Ridge," *Omaha Daily Bee*, 25 Aug. 1882; "A Black Cloud of Trouble Hovering over Pine Ridge Indian Agency," *St. Paul* (Minn.) *Daily Globe*, 29 Aug. 1882; "Indian Matters," *Salt Lake Daily Herald*, 29 Aug. 1882; Ostler, *Plains Sioux*, p. 205; Olson, *Red Cloud*, pp. 277–79.

26. Little Wound et al. to Commissioner of Indian Affairs, 18 Aug. 1882, in "A Black Cloud of Trouble." *See also* Olson, *Red Cloud*, p. 279.

27. "Indian Matters." *See also* Olson, *Red Cloud*, p. 281.

28. "Senator Dawes."

29. Pollock, quoted in "Inspector Pollock on Agent McGilly-cuddy," *Council Fire and Arbitrator* 6 (Sept. 1884):126.

30. Pollock, quoted in "Indian Pony Show," *Salt Lake Daily Herald*, 1 Sept. 1882.

31. Ibid.

32. "Agent McGillycuddy," *Omaha Daily Bee*, 25 Sept. 1882.

33. "What Red Cloud Asks For," *Council Fire and Arbitrator* 6 (Jan. 1883): 3; Moulton, *Valentine McGillycuddy*, p. 225.

34. "Indian Agent Investigated," *St. Paul Daily Globe*, 21 Dec. 1882.

35. Red Cloud, quoted in "Red Cloud's Ponies," *Salt Lake Daily Herald*, 31 Jan. 1883.

36. Ibid.; "Red Cloud Doesn't Want Cows," *St. Paul Daily Globe*, 26 Feb. 1883; Olson, *Red Cloud*, 233n69.

37. "Pine Ridge," *Omaha Daily Bee*, 25 Aug. 1882.

38. "After the Agent's Scalp," *Salt Lake Daily Herald*, 14 Mar. 1883; Ostler, *Plains Sioux*, pp. 207–9.

39. Ostler, *Plains Sioux*, pp. 207–9; Olson, *Red Cloud*, p. 297; Larson, *Red Cloud*, pp. 240–41; "How Others See It," *Council Fire and Arbitrator* 7 (Sept. 1884): 130; "Senator Dawes."

40. Red Cloud, quoted in "Red Cloud's Visit," *Washington* (D.C.) *Evening Critic*, 14 Mar. 1885.

41. Olson, *Red Cloud*, pp. 297–300.

42. Ibid., pp. 300–304.

43. Goodale, "Red Cloud and His Agent," *Southern Workman and Hampton School Record* 15 (Apr. 1886): 45–46. This version of

Goodale's article was reprinted from the *New York Evening Post*, but a date was not given.

44. Olson, *Red Cloud*, p. 304.

45. "McGillycuddy's Removal," *Omaha Daily Bee*, 19 May 1886.

CHAPTER 11: PINE RIDGE AGENCY, 1886–1893

1. Crook, quoted in "Sugar Talk," *St. Paul* (Minn.) *Daily Globe*, 19 June 1889.

2. Red Cloud, quoted in "Red Cloud on Capt. Bell," *McCook* (Nebr.) *Weekly Tribune*, 10 June 1886.

3. Bell, quoted in "Pine Ridge Indians," ibid., 21 Oct. 1886.

4. James C. Olson, *Red Cloud and the Sioux Problem* (Lincoln: University of Nebraska Press, 1965), pp. 307–8.

5. Robert M. Utley, *The Indian Frontier of the American West, 1846–1890* (Albuquerque: University of New Mexico Press, 1984), pp. 213–15; Arrell M. Gibson, "Indian Land Transfers," in *Handbook of North American Indians*, ed. William C. Sturtevant, 16 vols. (Washington, D.C.: Smithsonian Institution, 1978–), 4:226–27; Robert W. Larson, *Red Cloud: Warrior-Statesman of the Lakota Sioux* (Norman: University of Oklahoma Press, 1997), pp. 251–52. For the text of the Dawes Act, *see* Charles J. Kappler, ed., *Indian Affairs: Laws and Treaties*, 5 vols. (Washington, D.C.: Government Printing Office, 1904–1941), 1:33–36.

6. Olson, *Red Cloud*, p. 309; Larson, *Red Cloud*, p. 253.

7. "A Pow Wow of Braves," *Omaha Daily Bee*, 24 May 1888.

8. Olson, *Red Cloud*, pp. 310–11.

9. Jerome A. Greene, "The Sioux Land Commission of 1889: Prelude to Wounded Knee," *South Dakota History* 1 (Winter 1970): 45–46; Olson, *Red Cloud*, pp. 312–13. For the text of the 1889 Sioux Act, *see* Kappler, *Indian Affairs*, 1:328–39.

10. "Working Up Sentiment," *St. Paul Daily Globe*, 16 June 1889; Red Cloud, quoted in Olson, *Red Cloud*, p. 315.

11. Red Cloud, quoted in "Sugar Talk."

12. Utley, *Indian Frontier*, pp. 247–49; Larson, *Red Cloud*, p. 259.

13. Red Cloud, quoted in "Red Cloud Wants Wealth," *Casper* (Wyo.) *Weekly Mail*, 5 July 1889.

14. Wilds P. Richardson, "Some Observations upon the Sioux Campaign of 1890–1891," *Journal of the Military Service Institution of the United States* 18 (May 1896): 517.

15. Larson, *Red Cloud*, p. 261; Jeffrey Ostler, *The Plains Sioux and U.S. Colonialism from Lewis and Clark to Wounded Knee* (Cambridge, U.K.: Cambridge University Press, 2004), pp. 237–38; Utley, *Indian Frontier*, p. 249; "The Indians," *Sacramento* (Calif.) *Daily Record-Union*, 20 Dec. 1890.

16. Kappler, *Indian Affairs*, 1:56–58.

17. Smith D. Fry, "Indians Enter Protest," *Perrysburg* (Ohio) *Journal*, 19 June 1897.

18. Kappler, *Indian Affairs*, 1:105.

19. Edward Lazarus, *Black Hills, White Justice: The Sioux Nation versus the United States, 1775 to the Present* (New York: Harper-Collins, 1991), p. 125.

20. John D. McDermott, "Centennial Voices: The Tragedy at Wounded Knee," *South Dakota History* 22 (Winter 1990): 247–50; Ostler, *Plains Sioux*, pp. 237–39; Larson, *Red Cloud*, pp. 264–65.

21. Jack Goody, "Time: Social Organization," in *International Encyclopedia of the Social Sciences*, vol. 16, ed. David L. Sills (New York: The Macmillan Company and The Free Press, 1968), p. 41. For a lengthy discussion of this phenomenon, *see* Sylvia L. Thrupp, ed., *Millennial Dreams in Action: Studies in Revolutionary Religious Movements* (New York: Schocken Books, 1970).

22. Red Cloud, quoted in William S. E. Coleman, *Voices of Wounded Knee* (Lincoln: University of Nebraska Press, 2000), p. 238.

23. Robert M. Utley, *The Last Days of the Sioux Nation* (New Haven: Yale University Press, 1963), pp. 61–64, 71–72; Ostler, *Plains Sioux*, pp. 244–56.

24. Utley, *Last Days*, pp. 61–64, 71–72; Ostler, *Plains Sioux*, pp. 273–78, 291–93.

25. Utley, *Indian Frontier*, pp. 253–55, and *Last Days*, pp. 184–85; Larson, *Red Cloud*, pp. 270–71; Olson, *Red Cloud*, pp. 325–26.

26. Larson, *Red Cloud*, pp. 277–79; Utley, *Indian Frontier*, pp. 256–57. Estimates of Lakota deaths at Wounded Knee vary considerably, with some writers counting as many as three hundred. *See* Ostler, *Plains Sioux*, p. 345. For the best account of the Wounded Knee affair, *see* Jerome A. Greene, *American Carnage: Wounded Knee, 1890* (Norman: University of Oklahoma Press, 2014).

27. "Many Squaws Killed," *Pittsburg Dispatch*, 1 Jan. 1891.

28. Nicholas Black Elk and John G. Neihardt, *Black Elk Speaks* (Lincoln: University of Nebraska Press, 1979), p. 201.

29. "A Bloody Campaign," *Pittsburg Dispatch*, 1 Jan. 1891.

30. Olson, *Red Cloud*, pp. 323, 328.

31. Red Cloud to Bland, 12 Jan. 1891, in Olson, *Red Cloud*, p. 330.

32. "Red Cloud Coerced," *Omaha Daily Bee*, 4 Jan. 1891.

33. Charging Girl narrative, James H. Cook Papers, Agate Fossil Beds National Monument, Harrison, Nebr.

34. George R. Kolbenschlag, *A Whirlwind Passes: News Correspondents and the Sioux Indian Disturbances of 1890–1891* (Vermillion: University of South Dakota Press, 1990), p. 80.

35. Olson, *Red Cloud*, pp. 323, 328–29.

36. Taylor, quoted in *Eyewitnesses to the Indian Wars 1865–1890: The Long War for the Northern Plains*, ed. Peter Cozzens (Mechanicsburg, Pa.: Stackpole Books, 2004), pp. 611–14.

37. Fry, "Indians Enter Protest."

CHAPTER 12: LAST YEARS

1. Tilton, untitled poem, in "Red Cloud at a Reception," *Council Fire and Arbitrator* 6 (Jan. 1883): 6.

2. Willis Fletcher Johnson, *Life of Sitting Bull and History of the Indian War of 1890–91* (Edgewood, S.Dak.: Edgewood Publishing Co., 1891), p. 369.

3. George Bird Grinnell, "Red Cloud," ca. 1891, Folder 10, Box 2, Series 3, George Bird Grinnell Collection MO610, Nebraska State Historical Society, Lincoln.

4. Thisba Hutson Morgan, "Reminiscences of My Days in the Land of the Ogallala Sioux," *South Dakota Historical Collections* 29 (1958): 32; "The Passing of Red Cloud," *Denver Times*, 16 Apr. 1902.

5. Alfred J. Mokler, *History of Natrona County, Wyoming, 1888–1922* (Chicago: R. R. Donnelly & Sons, 1923), pp. 422–23; "Chief Red Cloud in the Natrona County Jail," *Wyoming Pioneer* 3 (Nov.-Dec. 1942): 19, 21–22; "Chief Red Cloud Once Arrested," *Casper* (Wyo.) *Times*, 1 May 1940.

6. Joy S. Kasson, *Buffalo Bill's Wild West: Celebrity, Memory, and Popular History* (New York: Hill and Wang, 2000), p. 212; Frank H. Goodyear III, *Red Cloud: Photographs of a Lakota Chief* (Lincoln: University of Nebraska Press, 2003), pp. 112–13, 119–21.

7. "Re-enact Custer: Realistic Sham Battle Fought at Pine Ridge," *Norfolk* (Nebr.) *Weekly News-Journal*, 1 Nov. 1902.

8. James R. Walker, *Lakota Belief and Ritual*, ed. Raymond J. DeMallie and Elaine A. Jahner (Lincoln: University of Nebraska Press, 1991), pp. 137–38.

9. Red Cloud, quoted in Jeffrey Ostler, *The Lakotas and the Black Hills: The Struggle for Sacred Ground* (New York: Viking Press, 2010), p. 129. *See also* Martin Luschei, *The Black Hills and the Indians: A Haven of Our Hopes* (San Luis Obispo, Calif.: Niobrara Press, 2007), pp. 135–36.

10. Edward Lazarus, *Black Hills, White Justice: The Sioux Nation versus the United States, 1775 to the Present* (New York: Harper-Collins, 1991), pp. 119–22.

11. "Aged Chief Red Cloud of the Sioux Indians Decides to Give Up Fight Against Government," *St. Louis Republic*, 15 Jan. 1905.

12. Warren K. Moorehead, *The American Indian in the United States: Period 1815–1914* (Andover, Mass.: Andover Press, 1914), p. 186.

13. "Old Red Cloud the Sioux Chief Dead," *Rapid City* (S.Dak.) *Journal*, 11 Dec. 1909; "Red Cloud Dead," *Valentine* (Nebr.) *Republican*, 17 Dec. 1909; James H. Cook, *Fifty Years on the Old Frontier as Cowboy, Hunter, Guide, Scout, and Ranchman* (Norman: University of Oklahoma Press, 1957), p. 184. Red Cloud was reportedly baptized a Catholic in the mid-1880s. Karl Markus Kreis, ed., *Lakotas, Black Robes, and Holy Women: German Reports from the Indian Missions in South Dakota, 1886–1900*, trans. Corinna Dally-Starna (Lincoln: University of Nebraska Press, 2000), p. 27. However, a 1974 government publication states that Red Cloud and his wife were baptized in 1908, taking the baptismal names John and Mary. *See* U.S., Department of the Interior, Bureau of Indian Affairs, *Famous Indians: A Collection of Short Biographies* (Washington, D.C.: Government Printing Office, 1974), p. 29.

14. "Red Cloud Dead," *Valentine Republican*, 17 Dec. 1909.

15. "Red Cloud, Sioux Chief Dead," *New York Times*, 11 Dec. 1909.

16. Valentine T. McGillycuddy to Walter M. Camp, 5 Dec. 1919, Walter M. Camp Papers, Little Bighorn Battlefield National Monument, Crow Agency, Mont.

17. Robert W. Larson, *Red Cloud: Warrior-Statesman of the Lakota Sioux* (Norman: University of Oklahoma Press, 1997), pp. 39–49.

18. McGillycuddy to Camp, 5 Dec. 1919.

19. Warren K. Moorehead, "The Passing of Red Cloud," *Transactions of the Kansas Historical Society, 1907–1908* (Topeka, 1908), p. 295.

20. "Red Cloud, The Indian Leader," *Omaha Weekly Herald*, 5 Apr. 1867.

21. "Frontier Sketches," *Denver Field and Farm*, 22 Oct. 1910.

CHAPTER 13: CONCLUSIONS: RED CLOUD'S MIND

1. James H. Cook, *Fifty Years on the Old Frontier as Cowboy, Hunter, Guide, Scout, and Ranchman* (Norman: University of Oklahoma Press, 1957), p. 184.

2. "Chief Red Cloud's Prayer," *Council Fire* 6 (June 1881): 88.

3. Red Cloud, quoted in James C. Olson, *Red Cloud and the Sioux Problem* (Lincoln: University of Nebraska Press, 1965), p. 113.

4. Sylvia Van Kirk, *Many Tender Ties: Women in Fur-Trade Society, 1670–1870* (Norman: University of Oklahoma Press, 1980), p. 9.

5. "1878 Speech of Red Cloud, Chief of the Ogalalla Sioux," *Council Fire* 2 (Mar. 1880): 37.

6. Red Cloud, quoted in Olson, *Red Cloud*, p. 112.

7. Red Cloud, quoted in Edmund B. Tuttle, *Three Years on the Plains: Observation of Indians, 1867–1870* (Norman: University of Oklahoma Press, 2003), pp. 182–83.

8. Doane Robinson, "The Education of Red Cloud," *South Dakota Historical Collections* 12 (1924): 174–75.

9. Red Cloud to Editor, 14 Jan. 1879, in "Strong Talk by a Straight-Tongued Sioux Chief," *Council Fire* 7 (Oct. 1884): 115.

10. Red Cloud, quoted in Walker, *Lakota Belief and Ritual*, p. 138.

11. R. Douglas Hurt, *Indian Agriculture in America: Prehistory to the Present* (Lawrence: University Press of Kansas, 1987), p. 134; John Dewey, "Interpretation of Savage Mind," *Psychological Review* 9 (1902): 217–30.

12. Red Cloud, quoted in Walker, *Lakota Belief and Ritual*, p. 137.

13. "Red Cloud Has His Hair Cut," *Washington Post*, 3 Sept. 1883.

14. Red Cloud and Plenty Bears to Editor, 24 Apr. 1884, in "From Red Cloud and Plenty Bears," *Council Fire* 7 (May 1884): 77.

15. Report of Agent Valentine T. McGillycuddy, Pine Ridge Agency, 1 Sept. 1881, in U.S., Department of the Interior, Office of Indian Affairs, *Annual Report of the Commissioner of Indian Affairs*

to the Secretary of the Interior for the Year 1881 (Washington, D.C.: Government Printing Office, 1881), pp. 46–47.

16. Woman's National Indian Association, "Facts Concerning the Indians," *Council Fire* 7 (Jan. 1884): 8.

17. Smith D. Fry, "Indians Enter Protest," *Perrysburg* (Ohio) *Journal*, 19 June 1897.

18. Woman's National Indian Association, "Facts Concerning the Indians," p. 8.

19. Red Cloud and Plenty Bears to Editor, 24 Apr. 1884.

20. "The Indian Problem," *St. Paul* (Minn.) *Daily Globe*, 26 Sept. 1885.

21. Red Cloud, quoted in T. R. Porter, "Famous Sioux Chief Red Cloud Lies on His Death Bed at Pine Ridge Agency," *Omaha World-Herald*, 9 Aug. 1903.

22. Edward Lazarus, *Black Hills, White Justice: The Sioux Nation versus the United States, 1775 to the Present* (New York: Harper-Collins, 1991), pp. 323–24, 375, 401; "Oglala Sioux to Consider Negotiating with Obama over Black Hills Claim," *Rapid City Journal*, 13 Nov. 2013.

Bibliography

ARCHIVAL COLLECTIONS

Agate Fossil Beds National Monument, Harrison, Nebr.
    James H. Cook Papers.
American Heritage Center, University of Wyoming,
        Laramie, Wyo.
    John Hunton Papers.
    Remi Nadeau Collection.
Beinecke Library, Yale University, New Haven, Conn.
    Coe Collection.
Billings Public Library, Billings, Mont.
    Indian Scrapbooks.
Dakota Indian Foundation, Chamberlain, S.Dak.
    Ella Deloria Collection.
Denver Public Library, Western History Room,
        Denver, Colo.
    William Collins Family Papers.
    Walter M. Camp Papers.
Fort Caspar Museum, Casper, Wyoming.
    Maxon Collection.
Fort Laramie National Historic Site, Wyoming.
    Fort Laramie National Historic Site Collections.
John D. McDermott Collection, Rapid City, S.Dak.
    Frank Aplan Correspondence.
    Writings of Alice Brown.
Lilly Library, Indiana University, Bloomington, Ind.
    Kenneth Hammer Collection.
Little Bighorn Battlefield National Monument,
        Crow Agency, Mont.
    Walter M. Camp Papers.
Library of Congress, Manuscript Division, Washington, D.C.
    Caleb Henry Carlton Papers.
    William J. Ghent Papers.
    Edward L. Hartz Papers.

Love Library, Special Collections, University of Nebraska,
    Lincoln, Nebr.
  Mari Sandoz Collection.
Montana Historical Society, Helena, Mont.
  James Brisbin Collection.
Nebraska State Historical Society, Lincoln, Nebr.
  George Bird Grinnell Collection.
  Addison E. Sheldon Collection.
  Eli S. Ricker Collection.
Newberry Library, Chicago, Ill.
  Elmo Scott Watson Papers.
North Dakota Historical Society, Bismarck, N.Dak.
  Lewis F. Crawford Papers.
University of Oklahoma Library, Western History Collections,
    Norman, Okla.
  W. S. Campbell Collection.
Rutherford B. Hayes Presidential Center, Fremont, Ohio.
  George R. Crook Papers.
South Dakota State Historical Society, State Archives Collection,
    Pierre, S.Dak.
  Biographical Files.
  Doane Robinson Papers.
  John R. Brennan Papers.
U.S. Army Military History Institute, Carlisle, Pa.
  Order of the Indian Wars Collection.
U.S. National Archives and Records Administration,
    Kansas City, Mo.
  Record Group 75, Records of the Office of Indian Affairs.
U.S. National Archives and Records Administration,
    Washington, D.C.
  Record Group 48, Records of the Secretary of the Interior.
  Record Group 75, Records of the Office of Indian Affairs.
  Record Group 94, Records of the Office of the Adjutant
    General.
  Record Group 123, Records of the United States Court of
    Claims.
  Record Group 204, Records of the Office of the Pardon
    Attorney, Department of Justice.
  Record Group 393, Records of United States Army Commands.

Wyoming State Archives, Cheyenne, Wyo.

Charles G. Coutant Collection.

GOVERNMENT DOCUMENTS

U.S. Department of the Interior. Bureau of Indian Affairs. *Famous Indians: A Collection of Short Biographies*. Washington, D.C.: Government Printing Office, 1974.

U.S. Department of the Interior. Office of Indian Affairs. *Annual Report of the Commissioner of Indian Affairs to the Secretary of the Interior*. Washington, D.C.: Government Printing Office, 1872, 1873, 1875–1877, 1880, 1881, 1884, 1913.

U.S. Department of the Interior. Office of Indian Affairs. *Report of the Special Commission Appointed to Investigate the Affairs of the Red Cloud Indian Agency, July, 1875, Together with the Testimony and Accompanying Documents*. Washington, D.C.: Government Printing Office, 1875.

U.S. Department of the Interior. Office of Indian Affairs. *Rules Governing the Court of Indian Offenses*. Washington, D.C.: Government Printing Office, 1883.

U.S. Department of the Interior. Office of Indian Affairs. "Census of Indians at Pine Ridge Agency, D.T.: Ogallala Sioux and Mixed Bloods." [1887.] In *Indian Census Rolls, 1885–1940*. Roll 363. National Archives Microfilm Publication M595. Washington, D.C.: National Archives and Records Administration, 1965.

U.S. Department of War. *Annual Report of the General of the Army to the Secretary of War*. Washington, D.C.: Government Printing Office, 1878.

BOOKS, THESES, AND DISSERTATIONS

Agonito, Joseph. *Lakota Portraits: Lives of the Legendary Plains People*. Guilford, Conn.: Twodot, 2011.

Albers, Patricia C., et. al. "The Home of the Bison: An Ethnographic and Ethnohistorical Study of Traditional Cultural Affiliations to Wind Cave National Park." Contracted study. National Park Service. 2003.

Alcorn, George Gregory. "The Powder River Country and the Sioux Treaty of 1868." Ph.D. diss., University of California, Santa Barbara, 1989.

Allen, Charles. *From Fort Laramie to Wounded Knee, In the West That Was.* Ed. and intro. Richard E. Jensen. Lincoln: University of Nebraska Press, 1997.

Athearn, Robert G. *William Tecumseh Sherman and the Settlement of the West.* Norman: University of Oklahoma Press, 1956.

Beard, Francis Birkhead. *Wyoming from Territorial Days to Present.* Chicago: The American Historical Society, 1933.

Berthrong, Donald J. *The Southern Cheyennes.* Norman: University of Oklahoma Press, 1963.

Bettelyoun, Susan Bordeaux, and Josephine Waggoner. *With My Own Eyes: A Lakota Woman Tells Her People's Story.* Ed. and intro. Emily Levine. Lincoln: University of Nebraska Press, 1988.

Birge, Julius C. *The Awakening of the Desert.* Boston: Richard C. Badger, 1912.

Black Elk, Nicholas, and John G. Neihardt. *Black Elk Speaks.* Lincoln: University of Nebraska Press, 1979.

Blaisdell, Bob, ed. *Great Speeches by Native Americans.* Mineola, N.Y.: Dover Publications, 2000.

Bourke, John G. *On the Border With Crook.* Lincoln: University of Nebraska Press, 1971.

Bray, Kingsley M. *Crazy Horse: A Lakota Life.* Norman: University of Oklahoma Press, 2006.

Brisbin, James S., ed. *Belden, the White Chief; or, Twelve Years among the Wild Indians of the Plains.* Cincinnati, Ohio: C. F. Vent, 1871.

Brown, Jennifer S. H. *Strangers in Blood: Fur Trade Families in Indian Country.* Vancouver: University of British Columbia Press, 1980; Norman: University of Oklahoma Press, 1990.

Brown, Joseph Epes. *Animals of the Sioux: Sacred Animals of the Oglala Sioux.* Rockport, Mass.: Element, 1992.

———. *The Spiritual Legacy of the American Indian: Commemorative Edition with Letters while Living with Black Elk.* Bloomington, Ind.: World Wisdom, 2007.

Buechel, Eugene. *Lakota Tales & Texts*, 2 vols. Chamberlain, S.Dak.: The Tipi Press, 1998.

Buecker, Thomas R. *Fort Robinson and the American West, 1874–1899.* Lincoln: Nebraska State Historical Society, 1999.

Calloway, Colin G. *One Vast Winter Count: The Native American West before Lewis and Clark*. Lincoln: University of Nebraska Press, 2003.

Carrington, Frances C. *My Army Life and the Fort Phil Kearney* [sic] *Massacre*. Philadelphia: Lippincott, 1910; Freeport, N.Y.: Books for Libraries, 1971; Boulder, Colo.: Pruett Publishing Co., 1990, intro. John D. McDermott.

Carrington, Margaret Irvin. *Absaraka, Home of the Crows*. Philadelphia: J. B. Lippincott, 1868.

Carrington, Margaret Irvin, and Henry B. Carrington. *Absaraka, Land of Massacre*. 3d ed. Philadelphia: J. B. Lippincott, 1878.

Coleman, William S. E. *Voices of Wounded Knee*. Lincoln: University of Nebraska Press, 2000.

Cook, James H. *Fifty Years on the Old Frontier, as Cowboy, Hunter, Guide, Scout, and Ranchman*. New Haven, Conn.: Yale University Press, 1923; Norman: University of Oklahoma Press, 1957, 1980.

Cozzens, Peter, ed. *Eyewitnesses to the Indian Wars 1865–1890: The Long War for the Northern Plains*. Mechanicsburg, Pa.: Stackpole Books, 2004.

Curtis, Edward S. *The North American Indian*. Vol. 3. Norwood, Mass.: By the Author, 1909.

DeMallie, Raymond J. *The Sixth Grandfather: Black Elk's Teachings Given to John G. Neihardt*. Lincoln: University of Nebraska Press, 1984.

Denig, Edwin. *Five Indian Tribes of the Upper Missouri: Sioux, Arickaras, Assiniboines, Crees, Crows*. Ed. John C. Ewers. Norman: University of Oklahoma Press, 1969.

Dodge, Richard Irving. *The Black Hills*. Minneapolis: Ross & Haines, 1965.

Doyle, Susan Badger. *Journeys to the Land of Gold: Emigrant Diaries from the Bozeman Trail, 1863–1866*. Vol. 2. Helena: Montana Historical Society Press, 2000.

Dunlay, Thomas W. *Wolves for the Blue Soldiers: Indian Scouts and Auxiliaries with the United States Army, 1860–90*. Lincoln: University of Nebraska Press, 1982.

Eastman, Charles A. *Indian Heroes & Great Chieftains*. Lincoln: University of Nebraska Press, 1991.

Ewers, John C., ed. *Indian Life on the Upper Missouri*. Norman: University of Oklahoma Press, 1968.

———. *Plains Indian History and Culture: Essays on Continuity and Change*. Norman: University of Oklahoma Press, 1997.

Frost, Lawrence A., ed. *With Custer in '74: James Calhoun's Diary of the Black Hills Expedition*. Provo, Utah: Brigham Young University Press, 1979.

Garrow, Carrie E., and Sarah Deer. *Tribal Criminal Law and Procedure*. Walnut Creek, Calif.: AltaMira Press, 2004.

Goodyear, Frank H. *Red Cloud: Photographs of a Lakota Chief*. Lincoln: University of Nebraska Press, 2003.

Gray, John S. *Centennial Campaign: The Sioux War of 1876*. Fort Collins, Colo.: The Old Army Press, 1976.

———. *Custer's Last Campaign: Mitch Boyer and the Little Bighorn Reconstructed*. Lincoln: University of Nebraska Press, 1991.

Greene, Jerome A. *American Carnage: Wounded Knee, 1890*. Norman: University of Oklahoma Press, 2014.

———. *Morning Star Dawn: The Powder River Expedition and the Northern Cheyennes, 1876*. Norman: University of Oklahoma Press, 2003.

Grinnell, George Bird. *The Fighting Cheyennes*. New York: Charles Scribner's Sons, 1915; Norman: University of Oklahoma Press, 1956.

Guernsey, Charles A. *Wyoming Cowboy Days*. New York: G. P. Putnam's Sons, 1936.

Hagan, William T. *Indian Police and Judges: Experiments in Acculturation and Control*. Lincoln: University of Nebraska Press, 1966.

Halaas, David Fridtjof, and Andrew E. Masich. *Halfbreed: The Remarkable True Story of George Bent–Caught Between the Worlds of the Indian and the White Man*. Cambridge, Mass.: Da Capo Press, 2004.

Hassrick, Royal B. *The Sioux: Life and Customs of a Warrior Society*. Norman: University of Oklahoma Press, 1964.

Hedren, Paul L. *Fort Laramie in 1876: Chronicle of a Frontier Post at War*. Lincoln: University of Nebraska Press, 1988.

Holt, O. H., comp. *Dakota*. Chicago: Rand McNally & Co., 1885.

Hurt, R. Douglas. *Indian Agriculture in America: Prehistory to the Present*. Lawrence: University Press of Kansas, 1987.

Hutton, Paul Andrew. *Phil Sheridan and His Army*. Lincoln: University of Nebraska Press, 1985.

Hyde, George E. *A Life of George Bent Written from His Letters*. Ed. Savoie Lottinville. Norman: University of Oklahoma Press, 1968.

———. *Red Cloud's Folk: A History of the Oglala Sioux Indians*. 1st ed. Norman: University of Oklahoma Press, 1937.

———. *Red Cloud's Folk: A History of the Oglala Sioux Indians*. Rev. ed. Norman: University of Oklahoma Press, 1975.

———. *A Sioux Chronicle*. Norman: University of Oklahoma Press, 1956.

———. *Spotted Tail's Folk: A History of the Brulé Sioux*. Norman: University of Oklahoma Press, 1960.

Hynes, William F. *Soldiers of the Frontier*. N.p.: Grand Council Fire of American Indians, 1943.

Jackson, Donald. *Custer's Gold: The United States Cavalry Expedition of 1874*. Lincoln: University of Nebraska Press, 1972.

Johnson, Willis Fletcher. *Life of Sitting Bull and History of the Indian War of 1890–91*. Edgewood, S.Dak.: Edgewood Publishing Co., 1891.

Kappler, Charles J., ed. *Indian Affairs: Laws and Treaties*. 5 vols. Washington, D.C.: Government Printing Office, 1904–1941.

———., comp. and ed. *Indian Treaties, 1778–1883*. Washington, D.C.: Government Printing Office, 1904; New York: Interland Publishing Co., 1972.

Kasson, Joy S. *Buffalo Bill's Wild West: Celebrity, Memory, and Popular History*. New York: Hill and Wang, 2000.

Kingsbury, George W. *History of Dakota Territory*, and George Martin Smith, *South Dakota: Its History and Its People*. 5 vols. Chicago: S. J. Clarke Publishing Co., 1915.

Kolbenschlag, George R. *A Whirlwind Passes: News Correspondents and the Sioux Indian Disturbances of 1890–1891*. Vermillion: University of South Dakota Press, 1990.

Kreis, Karl Markus, ed. *Lakotas, Black Robes, and Holy Women: German Reports from the Indian Missions in South Dakota, 1886–1900*. Trans. Corinna Dally-Starna. Lincoln: University of Nebraska Press, 2000.

Kroeker, Marvin E. *Great Plains Command: William Hazen in the Frontier West*. Norman: University of Oklahoma Press, 1976.

Lanham, Uri. *The Bone Hunters: The Heroic Age of Paleontology in the American West*. New York: Dover Publications, 1973.

Larson, Robert W. *Red Cloud, Warrior-Statesman of the Lakota Sioux*. Norman: University of Oklahoma Press, 1997.

Laubin, Reginald, and Gladys Laubin. *The Indian Tipi: Its History, Construction and Use*. New York: Ballantine Books, 1957.

Lazarus, Edward. *Black Hills, White Justice: The Sioux Nation versus the United States, 1775 to the Present*. New York: HarperCollins, 1991.

Lee, Bob, and Dick Williams. *Last Grass Frontier: The South Dakota Stock Grower Heritage*. South Dakota Stock Growers Association, 1964.

Leiker, James N., and Ramon Powers. *The Northern Cheyenne Exodus in History and Memory*. Norman: University of Oklahoma Press, 2011.

Luschei, Martin. *The Black Hills and the Indians: A Haven of Our Hopes*. San Luis Obispo, Calif.: Niobrara Press, 2007.

McChristian, Douglas C. *Fort Laramie: Military Bastion of the High Plains*. Norman, Okla.: Arthur H. Clark, 2008.

McDermott, John D. *Circle of Fire: The Indian War of 1865*. Mechanicsburg, Pa.: Stackpole Books, 2003.

———. *Frontier Crossroads: A History of Fort Caspar and the Upper Platte Crossing*. Casper, Wyo.: City of Casper, 1999.

———. *Gold Rush: The Black Hills Story*. Pierre: South Dakota State Historical Society Press, 2001.

———. *A Guide to the Indian Wars of the West*. Lincoln: University of Nebraska Press, 1998.

———. *Red Cloud's War: The Bozeman Trail, 1866–1868*. 2 vols. Norman, Okla.: Arthur H. Clark, 2010.

McGaa, Ed. *Crazy Horse and Chief Red Cloud: Warrior Chiefs–Teton Oglalas*. Minneapolis, Minn.: Four Directions Publishing, 2005.

McGillycuddy, Julia B. *McGillycuddy: Agent*. Stanford, Calif.: Stanford University Press, 1941. Reprinted as *Blood on the Moon: Valentine McGillycuddy and the Sioux*. Intro. James C. Olson. Lincoln: University of Nebraska Press, 1990.

McLaughlin, James. *My Friend the Indian*. Boston: Houghton Mifflin, 1910.

Maguire, H. N. *The Black Hills of Dakota: A Miniature History of Their Settlement, Resources, Populations, and Prospects*. Chicago: Jacob S. Gantz, 1879.

Mallery, Garrick. *Picture-Writing of the American Indians*. 2 vols. New York: Dover Publications, 1972.

Mokler, Alfred J. *Fort Caspar*. Casper, Wyo.: Prairie Publishing Co., 1939.

———. *History of Natrona County, Wyoming, 1888–1922*. Chicago: R. R. Donnelly & Sons, 1923.

Monnett, John. *Where A Hundred Soldiers Were Killed: The Struggle for the Powder River Country in 1868 and the Making of the Fetterman Myth*. Albuquerque: University of New Mexico Press, 2008.

Moorehead, Warren K. *The American Indian in the United States. Period 1850–1914*. Andover, Mass.: Andover Press, 1914.

Moulton, Candy V. *Valentine McGillycuddy: Army Surgeon, Agent to the Sioux*. Norman, Okla.: Arthur H. Clark, 2011.

Nadeau, Remi. *Fort Laramie and the Sioux*. Lincoln: University of Nebraska Press, 1967.

Olson, James C. *Red Cloud and the Sioux Problem*. Lincoln: University of Nebraska Press, 1965.

O'Meara, Walter. *Daughters of the Country: The Women of the Traders and Mountain Men*. New York: Harcourt, Brace, and World, 1968.

Ostler, Jeffrey. *The Lakotas and the Black Hills: The Struggle for Sacred Ground*. New York: Viking Press, 2010.

———. *The Plains Sioux and U.S. Colonialism from Lewis and Clark to Wounded Knee*. Cambridge, U.K.: Cambridge University Press, 2004.

Parker, Watson. *Gold in the Black Hills*. Norman: University of Oklahoma Press, 1966.

Paul, R. Eli., ed. *Autobiography of Red Cloud*. Helena: Montana Historical Society Press, 1997.

———. *Blue Water Creek and the First Sioux War, 1854–1856*. Norman: University of Oklahoma Press, 2004.

Powell, Peter J. *People of the Sacred Mountain: A History of the Northern Cheyenne Chiefs and Warrior Societies, 1830–1879*. Vol. 1. New York: Harper and Row, 1981.

Powers, Marla N. *Oglala Women: Myth, Ritual, and Reality*. Chicago: University of Chicago Press, 1986.

Powers, Thomas. *The Killing of Crazy Horse*. New York: Alfred A. Knopf, 2010.

Powers, William K. *Oglala Religion*. Lincoln: University of Nebraska Press, 1977.

Price, Catherine. *The Oglala People, 1841–1879: A Political History*. Lincoln: University of Nebraska Press, 1996.

Prucha, Francis Paul. *American Indian Treaties: The History of a Political Anomaly*. Berkeley: University of California Press, 1994.

Rambow, Charles. *Bear Butte: Journeys to the Sacred Mountain*. Sioux Falls, S.Dak.: By The Author, 2004.

Ricker, Eli S. *The Indian Interviews of Eli S. Ricker, 1903–1919*. Vol. 1 of *Voices of the American West*. Ed. Richard E. Jensen. Lincoln: University of Nebraska Press, 2005.

Riggs, Stephen R. *Dakota Grammar, Texts, and Ethnography*. Ed. James Owen Dorsey. 1893. Reprint, Marvin, S.Dak.: American Indian Culture Research Center, 1977.

Robinson, Doane. *A History of the Dakota or Sioux Indians*. Minneapolis: Ross & Haines, 1956.

Sandoz, Mari. *Crazy Horse, the Strange Man of the Oglalas*. New York: Hastings House, Inc., 1942; Lincoln: University of Nebraska Press, 1961.

———. *The Great Council*, ed. Caroline Sandoz Pifer. Crawford, Nebr.: Cottonwood Press, 1982.

Scott, Hugh Lennox. *Some Memories of a Soldier*. New York: The Century Company, 1928.

Seymour, Flora Warren. *Indian Agents of the Old Frontier*. New York: F. Appleton-Century Company, 1941.

Simon, John Y., ed. *The Papers of Ulysses S. Grant*. 32 vols. Carbondale: Southern Illinois University Press, 1967–2012.

Standing Bear, Luther. *Land of the Spotted Eagle*. Lincoln: University of Nebraska Press, 1978.

Stewart, Edgar I. *Custer's Luck*. Norman: University of Oklahoma Press, 1955.

Sundberg, Wayne. *The Story of Camp and Fort Collins*. Fort Collins, Colo., n. p., 1985.

Sundstrom, Linea. *Storied Stone: Indian Rock Art in the Black Hills Country*. Norman: University of Oklahoma Press, 2004.

Taopi and His Friends or the Indians' Wrongs and Rights.
Philadelphia: Claxton, Remsen & Haffelfinger, 1869.

Thorne, Tanis C. *The Many Hands of My Relations: French and Indians on the Lower Missouri.* Columbia: University of Missouri Press, 1996.

Thrupp, Sylvia L., ed. *Millennial Dreams in Action: Studies in Revolutionary Religious Movements.* New York: Schocken Books, 1970.

Tuchner, Lesta V., and James D. McLaird. *The Black Hills Expedition of 1875.* Mitchell, S.Dak.: Dakota Wesleyan University Press, 1975.

Tuttle, Edmund B. *Three Years on the Plains: Observations of Indians, 1867–1870.* Norman: University of Oklahoma Press, 2003.

Utley, Robert M. *The Indian Frontier of the American West, 1846–1890.* 1st ed. Albuquerque: University of New Mexico Press, 1984.

———. *The Indian Frontier of the American West, 1846–1890.* Rev. ed. Albuquerque: University of New Mexico Press, 2003.

———. *The Last Days of the Sioux Nation.* New Haven, Conn.: Yale University Press, 1963.

Van Kirk, Sylvia. *Many Tender Ties: Women in Fur-Trade Society, 1670–1870.* Norman: University of Oklahoma Press, 1980.

Vaughn, J. W. *The Battle of Platte Bridge.* Norman: University of Oklahoma Press, 1963.

Vestal, Stanley. *Warpath: The True Story of the Fighting Sioux Told in a Biography of White Bull.* Boston: Houghton Mifflin, 1934.

———. *Warpath and Council Fire: The Plains Indian Struggle for Survival in War and in Diplomacy, 1851–1891.* New York: Random House, 1948.

Walker, James R. *Lakota Ritual and Belief.* Ed. Raymond J. DeMallie and Elaine A. Jahner. Lincoln: University of Nebraska Press, 1991.

———. *Lakota Society.* Ed. Raymond J. DeMallie. Lincoln: University of Nebraska Press, 1982.

ARTICLES

Adams, Donald K. "The Journal of Ada A. Vogdes, 1868–71." *Montana, the Magazine of Western History* 13 (July 1963): 2–17.

Agonito, Joseph. "The Art of Plains Indian Warfare." *Order of the Indian Wars Journal* 1 (Winter 1980): 1–21.

Allen, Charles W. "Red Cloud, Chief of the Sioux." *The Hesperian* 1 (Jan. 1896): 211–16.

———. "Red Cloud and the U.S. Flag." *Nebraska History* 21 (Oct.–Dec. 1940): 293–304.

Andrist, Ralph K. "Red Cloud of the Oglala Sioux." *The Westerners New York Posse Brand Book* 15 (1968): 5–6.

Behrens, Jo Lea Wetherilt. "In Defense of 'Poor Lo': National Indian Defense Association and *Council Fire*'s Advocacy for Sioux Land Rights." *South Dakota History* 24 (Fall/Winter 1994): 153–73.

Berthrong, Donald J. "Changing Concepts: The Indians Learn about the 'Long Knives' and Settlers, 1849–1890s." In *Red Men and Hat Wearers: Viewpoints in Indian History*. Ed. Daniel Tyler. Fort Collins: Colorado State University Press, 1976. Pp. 47–62.

Bland, Thomas A. "McGillycuddy's Letter to the Press." *Council Fire* 7 (Oct. 1884): 146.

———. "Off for Dakota." *Council Fire* 7 (June 1884): 103.

Bray, Kingsley M. "Lone Horn's Peace: A New View of Sioux-Crow Relations, 1851–1858." *Nebraska History* 66 (Spring 1985): 28–47.

———. "The Oglala Lakota and the Establishment of Fort Laramie." *Museum of the Fur Trade Quarterly* 36 (Winter 2000): 3–18.

Brown, Joseph Epes. "The Unlikely Associates: A Study in Oglala Sioux Magic and Metaphysic." *Studies in Comparative Religion* 4 (Summer 1970): 1–9.

Buecker, Thomas R. "'Distance Lends Enchantment to the View': A Diary of the 1874 Black Hills Expedition." In *Beyond Mount Rushmore*. Ed. and intro. Mary C. Kopco. Pierre: South Dakota State Historical Society Press, 2010. Pp. 9–13.

Calloway, Colin G. "The Only Way Open to Us: The Crow Struggle for Survival in the Nineteenth Century." *North Dakota History* 33 (Summer 1980): 24–34.

"Chief Red Cloud in the Natrona County Jail." *The Wyoming Pioneer: An Historical Publication* 3 (Nov.–Dec. 1942): 19, 21–22.

"Chief Red Cloud's Prayer." *Council Fire* 6 (June 1881): 88.

Clow, Richmond L. "The Sioux Nation and Indian Territory: The Attempted Removal of 1876." *South Dakota History* 6 (Fall 1976): 456–73.

Collins, Charles. "A Massacre Averted: How Spotted Tail and His Associates Saved the Lives of the Commissioners." *Council Fire* 4 (Sept. 1881): 135.

Conger, A. L. "The Function of Military History." *Mississippi Valley Historical Review* 3 (Sept. 1916): 161–71.

Cook, James H. "The Art of Fighting Indians." *American Mercury* 23 (June 1931): 170–79.

Daniels, Robert E. "Cultural Identities among the Oglala Sioux." In *The Modern Sioux: Social Systems and Reservation Culture*. Ed. Ethel Nurge. Lincoln: University of Nebraska Press, 1970. Pp. 198–245.

DeMallie, Raymond J. "Lakota Belief and Ritual in the Nineteenth Century." In *Sioux Indian Religion: Tradition and Innovation*. Ed. Raymond J. DeMallie and Douglas R. Parks. Norman: University of Oklahoma Press, 1987. Pp. 25–44.

———. "Male and Female in Traditional Lakota Culture." In *The Hidden Half: Studies of Plains Indian Women*. Ed. Patricia Albers and Beatrice Medicine. Lanham, Md.: University Press of America, 1983. Pp. 237–65.

———. "Sioux until 1850." In *Handbook of North American Indians*. Vol. 13. Ed. William C. Sturtevant. Washington, D.C.: Smithsonian Institution, 2001. Pp. 718–60.

———. "Teton." In *Handbook of North American Indians*. Vol. 13. Ed. William C. Sturtevant. Washington, D.C.: Smithsonian Institution, 2001. Pp. 794–820.

———. "Touching the Pen." In *Ethnicity on the Great Plains*. Ed. Frederick C. Luebke. Lincoln: University of Nebraska Press, 1980. Pp. 38–53.

DeMallie, Raymond J., and Douglas A. Parks. "Plains Indian Warfare." In *The People of the Buffalo: Plains Indians of North America, Military Art, Warfare and Change*. Ed. Colin F. Taylor and Hugh A. Dempsey. Wyk auf Foehr, Germany: Tatanka Press, 2003. Pp. 66–76.

Dewey, John. "Interpretation of Savage Mind." *Psychological Review* 9 (1902): 217–30.

Dickson, Ephriam D., III. "Dakota Resources: The Sitting Bull Surrender Census, Standing Rock Agency, 1881." *South Dakota History* 40 (Summer 2010): 163–96.

Dorsey, J. Owen. "The Social Organization of the Siouan Tribes." *Journal of American Folklore* 4 (1896): 262–63.

Doyle, Susan Badger. "The Bozeman Trail, 1863–1868." *Overland Journal* 20 (Spring 2002): 2–17.

———. "Indian Perspectives of the Bozeman Trail." *Montana, the Magazine of Western History* 40 (Winter 1990): 56–65.

"1878 Speech of Red Cloud, Chief of the Ogalalla Sioux." *Council Fire* 2 (Mar. 1880): 37; republished in *The Council Fire and Arbitrator* 7 (Oct. 1888): 140–41.

Ellis, Mark R. "Reservation Akicitas: The Pine Ridge Indian Police, 1879–1885." *South Dakota History* 29 (Fall 1999): 185–210.

Fletcher, Alice C. "The Elk Mystery or Festival, Ogallala Sioux." In *Sixteenth and Seventeenth Annual Reports of the Trustees of the Peabody Museum of American Archaeology and Ethnology.* Cambridge, Mass.: Peabody Museum, 1884. Pp. 276–88.

"The Fraud of the Sioux Commission Exposed." *Council Fire* 6 (October 1883): 138–39.

"From Red Cloud and Plenty Bears." *Council Fire* 7 (May 1884): 77.

Gibson, Arrell M. "Indian Land Transfers." In *Handbook of North American Indians.* Vol. 4. Ed. William C. Sturtevant. Washington, D.C.: Smithsonian Institution, 1988. Pp. 211–29.

Goldfrank, Esther S. "Historic Change and Social Character: A Study of the Teton Dakota." *American Anthropologist* 45 (Jan.-Mar. 1943): 67–83.

Goody, Jack. "Time: Social Organization." In *International Encyclopedia of the Social Sciences.* Vol. 16. Ed. David L. Sills. New York: The Macmillan Company and The Free Press, 1968. Pp. 41.

"The Great Chief Red Cloud." *Indian School Journal* 10 (Feb. 1910): 41–42.

Greene, Jerome A. "The Sioux Land Commission of 1889: Prelude to Wounded Knee." *South Dakota History* 1 (Winter 1970): 41–72.

Grinnell, George Bird. "Tenure of Land Among the Indians." *American Anthropologist* 9 (Jan-May 1907): 1–11.

Hagan, William T. "Justifying Dispossession of the Indian." In *American Indian Environments: Ecological Issues in Native American History.* Ed. Christopher Vecsey and Robert W.

Venables. Syracuse, N.Y.: Syracuse University Press, 1980.
Pp. 65–80.

Hassrick, Royal B. "The Sioux Indians." *The Westerners Brand Book* 17 (1961): 49–64.

Hayden, Ferdinand V. "Address on the Black Hills." *Proceedings of the American Philosophical Society* 10 (1869): 322–26.

Hinman, Eleanor. "Oglala Sources of the Life of Crazy Horse." *Nebraska History* 57 (1976): 1–51.

"How Others See It." *Council Fire* 6 (September 1884): 130.

Jones, Brian. "John Richard, Jr. and the Killing at Fetterman." *Annals of Wyoming* 43 (Fall 1971): 237–58.

Keller, Robert H., Jr. "A Puritan at Alder Gulch and the Great Salt Lake: Rev. Jonathan Blanchard's Letters from the West, 1864." *Montana, the Magazine of Western History* 36 (Summer 1975): 62–75.

Larson, Robert W. "Lakota Leaders and Government Agents: A Story of Changing Relationships." *Nebraska History* 82 (Summer 2001): 47–57.

———. "Red Cloud: A Warrior-Statesman's Influence." *Greasy Grass* 24 (2008): 15–20.

McDermott, John D. "A Dedication to the Memory of George Hyde, 1882–1968." *Arizona and the West* 17 (Summer 1975): 103–6.

———. "Centennial Voices: The Tragedy at Wounded Knee." *South Dakota History* 20 (Winter 1990): 245–98.

———. "John Baptiste Richard." In *The Mountain Men and the Fur Trade of the Far West*. Vol. 2. Ed. LeRoy R. Hafen. Glendale, Calif.: Arthur H. Clark, 1965. Pp. 289–303.

———. "The Short and Controversial Life of William Judd Fetterman." *Annals of Wyoming* 63 (Spring 1991): 49–53.

———. "'We had a terribly hard time letting them go': The Battles of Mud Springs & Rush Creek, February, 1865." *Nebraska History* 77 (Summer 1996): 78–88.

McGee, W. J. "The Sioux Indians: A Preliminary Sketch." In *The Sioux Indians: A Socio-Ethnological History*. Ed. John M. Carroll. Intro. John F. Bryde. New York: Sol Lewis, 1973. Pp. 5–52.

McGillycuddy, Valentine T. "First Survey of the Black Hills." *Motor Travel* 20 (Oct. 1928): 18.

Manypenny, George W. "Red Cloud, Chief of the Ogallalla Sioux." *Council Fire* 7 (Dec. 1884): 174–77.

Means, Jeffrey D. "'Indians Shall Do Things in Common': Oglala Lakota Identity and Cattle-Raising." *Montana, the Magazine of Western History* 61 (Autumn 2011): 3–21.

Mekeel, H. Scudder. "A Discussion of Culture Change As Illustrated by Material from a Teton-Dakota Community." *American Anthropologist* 34 (Apr.-June 1932): 274–85.

Mokler, Verne. "Story of Caspar Collins: Oregon Trail Trek No. Three." *Annals of Wyoming* 28 (Oct. 1956): 180–87.

Moorehead, Warren King. "The Passing of Red Cloud." *Transactions of the Kansas Historical Society, 1907–1908* (Topeka, 1908): 295–311.

Morgan, Thisba Hutson. "Reminiscences of My Days in the Land of the Ogallala Sioux." *South Dakota Historical Collections* 29 (1958): 21–62.

Murphy, James C. "The Place of the Northern Arapahoes in Relations between the U.S. and the Indians of the Plains, 1851–1879." Parts 1 and 2. *Annals of Wyoming* 41 (Apr. 1969): 33–61; (Oct. 1969): 203–59.

Newcomb, W. W., Jr. "A Re-examination of the Causes of Plains Warfare." *American Anthropologist* 52 (July-Sept. 1950): 317–30.

Parker, Watson. "The Black Hills Controversy." Paper presented at the annual meeting of the Dakota History Conference, Apr. 1984.

Paul, R. Eli. "Dakota Resources: The Investigation of Special Agent Cooper and Property Damage Claims in the Winter of 1890–1891." *South Dakota History* 24 (Fall-Winter 1994): 212–35.

———. "Recovering Red Cloud's Autobiography: The Strange Odyssey of a Chief's Personal Narrative." *Montana, the Magazine of Western History* 44 (Summer 1994): 2–17.

Pennington, Robert. "An Analysis of the Political Structure of the Teton-Dakota Indian Tribe of North America." *North Dakota History* 20 (July 1953): 143–56.

Pollock, William J. "Inspector Pollock on Agent McGillycuddy," *Council Fire* 6 (Sept. 1884): 126–30.

Pond, Gideon H. "Dakota Superstitions." *Collections of the Minnesota Historical Society* 2 (1860–1867): 215–18.

Potter, Reuben M. "The Red Man's God." *Journal of the Military Service Institute* 7 (Mar. 1886): 621–71.

"Red Cloud, His Prayer." *Wyoming Churchman* 2 (Jan. 1912): 9.

"Red Cloud Becomes Civilized." *Indian School Journal* 5 (Feb. 1905): 17–19.

Richardson, Wilds P. "Some Observations upon the Sioux Campaign of 1890–1891." *Journal of the Military Service Institution* 18 (May 1896): 512–31.

Ripich, Carol A. "Joseph Wham and the Red Cloud Agency." *Arizona and the West* 12 (Winter 1970): 325–38.

Robinson, Doane. "The Education of Red Cloud." *South Dakota Historical Collections* 12 (1924): 156–78.

———. "The Sioux of the Dakotas." *Home Geographic Monthly* 2 (Nov. 1932): 8–12.

"Sioux Women at Home." *The Illustrated American* (31 Jan. 1891): 481–86.

Sleeper-Smith, Susan. "Furs and Female Kin Networks: The World of Marie Madeleine Réaume L'archeveque Chevalier." In *New Faces of the Fur Trade: Selected Papers of the Seventh North American Fur Trade Conference, Halifax, Nova Scotia, 1995.* Ed. Jo-Anne Fiske, Susan Sleeper-Smith, and William Wicken. Lansing: Michigan State University Press, 1998. Pp. 53–72.

Smith, Sherry L. "Officers' Wives, Indians, and the Indian Wars." *Order of the Indian Wars Journal* 1 (Winter 1980): 32–50.

Spring, Agnes Wright. "The Founding of Fort Collins, United States Military Post." *Colorado Magazine* 10 (1933): 47–55.

Sundstrom, Linea. "Cross-Cultural Transference of the Sacred Geography of the Black Hills." *World Archaeology* 28 (Oct. 1996): 177–90.

———. "The Sacred Black Hills: An Ethnohistorical Review." *Great Plains Quarterly* 17 (Summer/Fall 1997): 185–212.

Twitchell, Phillip G., ed. "Camp Robinson Letters of Angeline Johnson, 1876–1879." *Nebraska History* 77 (Summer 1996): 89–95.

Utley, Robert M. "The Bozeman Trail before John Bozeman: A Busy Land." *Montana, the Magazine of Western History* 53 (Summer 2003): 20–31.

———. "The Celebrated Peace Policy of General Grant." *North Dakota History* 20 (July 1953): 121–42.

Vecsey, Christopher. "American Indian Environmental Religions." In *American Indian Environments: Ecological Issues in Native American History*. Ed. Christopher Vecsey and Robert W. Venables. Syracuse, N.Y.: Syracuse University Press, 1980. Pp. 1–37.

Wade, Arthur P. "The Military Command Structure: The Great Plains, 1853–1891." *Journal of the West* 15 (July 1976): 5–22.

Walker, James R. "The Sundance and Other Ceremonies of the Oglala Division of the Teton Dakota." *Anthropological Papers of the American Museum of Natural History* 16 (1917): 50–221.

Welsh, Herbert. "The Meaning of the Dakota Outbreak." *Scribner's Magazine* 9 (Apr. 1891): 439–52.

"Why They Sustain McGillycuddy," *Council Fire* 7 (Oct. 1884): 147–48.

Wilson, E. P. "The Story of the Oglala and Brule Sioux in the Pine Ridge Country of Northwest Nebraska in the Middle Seventies." *Nebraska History* 21 (Oct.-Dec. 1940): 259–74.

Wissler, Clark. "Depression and Revolt." *Natural History* 41 (Feb. 1938): 108–12.

———. "Societies and Ceremonial Associations in the Oglala Division of the Teton-Dakota." *Anthropological Papers of the American Museum of Natural History* 11 (1912): 1–99.

Woman's National Indian Association. "Facts Concerning the Indians." *Council Fire* 7 (Jan. 1884): 8–9.

NEWSPAPERS

Army and Navy Journal, 1868.

Army and Navy Register, 1891.

Baltimore Sun, 1876.

Bismarck (D.T.) *Tribune*, 1874.

Butte (Mont.) *Miner*, 1876.

Casper (Wyo.) *Times*, 1940.

Central City (Colo.) *Miner's Register*, 1865.

Cheyenne (Wyo.) *Daily Leader*, 1868–1890.

Cheyenne (Wyo.) *Daily Sun-Leader*, 1897.

Cheyenne (Wyo.) *State Tribune*, 1944.

Chicago Daily Sun, 1897.
Chicago Inter-Ocean, 1876.
Chicago Tribune, 1875.
Denver Daily Rocky Mountain News, 1866.
Denver Farm and Fireside, 1922.
Denver Field & Farm, 1895.
Denver Times, 1902.
Harvard Independent, 1910.
McCook (Nebr.) *Tribune*, 1886.
Maysville (Kans.) *Daily Evening Bulletin*, 1885.
Nebraska State Journal (Lincoln), 1909–1917.
New York Graphic, 1877.
New York Herald, 1875–1877.
New York Sun, 1891–1914.
New York Times, 1867–1909.
New York Tribune, 1908.
New York World, 1874.
Norfolk (Nebr.) *Weekly News-Journal*, 1902.
Omaha Daily Bee, 1875–1891.
Omaha Daily World, 1875.
Omaha Tribune and Republican, 1872.
Omaha Weekly Herald, 1867.
Omaha World-Herald, 1903, 1927.
Perrysburg (Ohio) *Journal*, 1897.
Philadelphia Inquirer, 1874.
Pittsburg Dispatch, 1891.
Rapid City (S.Dak.) *Journal*, 1909, 2013.
Richmond (Ky.) *Climax*, 1890.
Sacramento (Calif.) *Daily Record-Union*, 1890.
St. Louis Republic, 1905.
St. Paul and Minneapolis Pioneer Press, 1876–1886.
St. Paul (Minn.) *Daily Globe*, 1882–1889.
Salt Lake Daily Union Vedette, 1866.
Salt Lake Herald, 1882–1883.
Sheridan (Wyo.) *Daily Enterprise*, 1909.
Spearfish (S.Dak.) *Daily Bulletin*, 1890.
Valentine (Nebr.) *Republican*, 1909.
Washington (D.C.) *Evening Critic*, 1885.

Washington (D.C.) *Evening Star*, 1876.

Washington (D.C.) *Post*, 1883–1889.

Wyoming Weekly Leader (Cheyenne), 1869.

Index